THE GREAT DIVERSITY

THE GREAT DIVERSITY

Trajectories of Asian Development

edited by:

Camilla Brautaset

Christopher M. Dent

Wageningen Academic
Publishers

ISBN: 978-90-8686-233-7
e-ISBN: 978-90-8686-786-8
DOI: 10.3920/978-90-8686-786-8

Photos cover: Mike Parnwell

First published, 2014

Preface

Asia is on our minds these days. In the West, Asia's recent development has provoked apprehension, admiration, adaptation – and a frequent arising of catch-22 situations. For instance, whilst the media home in on how cheap Asian products pose a threat to workplaces elsewhere, at the same time we are told how important it is to maintain high economic growth figures in Asia in general and China in particular in order to keep the global economy afloat and progressing. Furthermore, economists continue to wrangle over the causes and sustainability of Asia's dynamic economic development. Certainly, Asia is expected by most to have more impact on the future global economy than any other region. Political scientists from the West question whether the increased geopolitical importance of Asia is a threat to the liberal world order or not. Moreover, political economists and anthropologists ask how new global and Asian developments affect human behaviour, the environment and the structures of our societies. At the same time, the influx of Asian popular culture in the West is becoming part of the mainstream. During the summer of 2013, the Japan Expo in Paris and the Hyper Japan in London were the largest pop-cultural events in Europe.

Though historians appear to have agreed to disagree on Asia's global importance in the past, there can be little doubt that this region – so rich in resources, culture and history – holds one of the master keys to global development in the 21st century. Still, in order to gain a fuller comprehension of the complexities of Asian development it is imperative to move beyond academic compartmentalisation and institutional vested knowledge. This was exactly the ambition behind a workshop held in Bergen in September 2012. Scholars from the universities of Bergen and Leeds representing a wide range of sub-disciplines within the social sciences and the humanities gathered in order to exchange views and approaches and to explore whether there was a common ground in order to pursue closer co-operation in the future.

This book is a product of this workshop. The title – *The Great Diversity. Asian Trajectories of Development* – signals an ambition to capture some of the complexities that characterise the Asian paths of development in the past, present and future. This anthology is as diverse as the developments it sets out to describe and offers a taste of on-going research on Asian development at the University of Bergen and the University of Leeds. Both these universities are part of the World Universities Network (WUN). We are grateful for the support from both our home universities. In particular, we would like to thank Bjørn Helge Andersen and the Department of Research Management at the University of Bergen and Matthias Kaiser and the Centre for the Study of the Humanities and the Sciences at the University of Bergen for their generous co-funding of the conference. Most of the funding has, however, come through the Norwegian Research Council and Project Number 205553, which is hosted by Department of Archaeology, History, Cultural Studies and Religion at the University of Bergen. It is hoped this project will lay down a strong foundation stone on which to build future expansive collaboration on Asia-related research between Bergen and Leeds.

Bergen, August 2013
Camilla Brautaset and Christopher M. Dent

Table of contents

Chapter 1: Introduction: diversity and development in Asia

Christopher M. Dent and Camilla Brautaset

The Universities of Bergen and Leeds are both member institutions of the World Universities Network (WUN). This book brings together Asia specialists from both universities in a collaborative venture to explore different perspectives on how we may understand development, and more specifically the development trajectories of Asia, arguably the most dynamic and certainly the most diverse part of our world. The 'great diversity' of Asia reminds us that there is no one singular 'truth' on understanding development, or universal model on prescribing future paths of development. The importance of locality has not diminished despite deepening globalisation in the modern era. Accepting the prevalence of diversity does not mean we cannot learn certain lessons of development from each other, and this applies from an academic disciplinary perspective as well as those from the same disciplinary background studying different parts of Asia. The scholars from Bergen and Leeds assembled for this book come from a variety of backgrounds: history, economics, geography, politics, sociology, anthropology, environmental science, international political economy, religious studies and philosophy.

The term 'development' itself is highly contested, and there exist multiple narratives and discourses on the subject across and within different academic disciplines. To many, the notion of development is strongly linked to ideas arising from the European 'enlightenment' concerning the progressive or sustained improvement of individual and societal welfare. This, it has been long argued, is the overarching, logical and scientific pursuit of human civilisation broadly conceived. However, interpretations of improved welfare, and thereby 'progressive' development, have changed over time and have moreover become increasingly contested. Many also prescribe to an 'anti-development' viewpoint by questioning whether development in itself is laudable objective, for example where it leads to ecological damage or collapse.

This book does not seek to define development, or prescribe a particular method of understanding it in an Asian context. Rather, it presents a number of works that in their own way touch on the subject of development, and lays bare the inherent diversity of development as an idea, practice and experience. It is up to the reader to reflect on how the evidence and arguments presented in each chapter resonates, or not, on their own understanding of development. It is too broad a subject or concept to be confined by set analytical parameters that aim to unify explanations

C. Brautaset and C.M. Dent (eds.) **THE GREAT DIVERSITY**
DOI 10.3920/978-90-8686-786-8_1, © Wageningen Academic Publishers 2014

for what is essentially a highly diverse phenomenon. What is arguably more useful is to consider the meaning of development from a range of scholarly perspectives and look for the connections between them, thus help foster cross-disciplinary learning on the matter.

Structure of this book

The book is structured along the following broad lines. Chapters 2 and 3 present historic perspectives on Asian development and Chinese-Western connections in particular, Chapters 4 to 8 look at various social, cultural and religious themes, and Chapters 9 to 12 examine different aspects of economic and business development.

In Chapter 2 'Merchants and missionaries: connecting China, Norway and beyond', Camilla Brautaset examines the development of the transnational links between China and Norway during the late 19th and early 20th centuries, reminding us of the historic dimension of China's engagement with the forces of globalisation. The relationship between China and the West during this period has been at the epicentre of several overlapping – and interacting – academic discourses over the past decades. Brautaset argues how global history has become a *genre* of comparisons and connectedness in time and space, whilst globalisation studies have a *thematic* focus on the dynamics of macro integration. In her chapter, Brautaset also outlines the motivation, ambition and research design of the ongoing project 'Merchants and missionaries: Norwegian encounters with China in a transnational perspective, 1890-1937', which she is leading. This project is multidisciplinary in approach, and pays especially heed to the theoretical and practical applications of networks and mobility in historical studies. Brautaset claims that historical network studies represent a key to further our understanding of both geographical and social mobility. She suggests that migration theories may offer what may be a missing link in existing studies of the mobility of merchants as well as missionaries. By combining qualitative and quantitative sources, she suggests that agents acted according to communicative as well as calculative rationalities, and that they were immensely adaptable in their behaviour. Hence, networks appear to define a room of manoeuvre, but not necessarily a pre-destined pattern of behaviour.

This is followed in Chapter 3 'Opening the gates: Maritime Customs Service (1854-1949) in the Chinese contemporary context' by Olga Medvedeva's study on the development of China's Maritime Customs Service from the mid-19th century to the mid-20th century. The origins for this service can be dated back to the 7th century, and Medvedeva contends that its evolution was closely connected to China's changing attitudes on international trade and the threat from foreign maritime powers. She notes that by employing people from over 20 different nationalities, the Chinese Maritime Customs Service became an important cosmopolitan organization in the country, and a key agent of China's internationalisation during the study period. The chapter pays special attention to how the development of the organisation was perceived by Chinese scholars at the time. Medvedeva observes that these scholars viewed the employment of foreigners in China's customs service was seen as generally positive from a technical perspective, yet also rousing national sentiments that eventually led to moves towards a more protectionist stance on trade, and a more defensive stance of China's national interests.

Continuing in an historical vein, Karina Hestad Skeie in Chapter 4 'The world wide missionary web: transnational mission networks' influence on the 19th and 20th Century Norwegian China missions' studies how the Protestant mission movement in the 19th and 20th centuries mobilized Christians across continents and national borders, and created transnational networks of people, institutions, ideas, money and methods, with a particular focus on Norwegian and other Scandinavian missionaries working in China. Traditional mission studies tend to follow national and denominational boundaries. This chapter, however, addresses the extent to which individuals and organizations were influenced by each other through networks and transnational collective actions. The case studies investigated here, show how the missionary networks were critically dependent on mutual trust in order to function properly, and how the mission societies strived to minimise as well as to control risk. Moreover, Hestad Skeie examines the importance of representation and meaning-production, as well as the continued importance of the 'national' even as it was transgressed and challenged. She argues that the many Norwegian China missions reflect the various ways transnational impulses were appropriated and transformed locally.

From quite a different social, cultural and religious perspective, Michael Parnwell in Chapter 5 'Regional resonances: Vietnamese whale worship in comparative perspective' studies the veneration of whales (and other cetaceans) by fishing communities in central and southern Vietnam, which can be traced back at least 250 years. This practice centres on elaborate funerals organised for whales that die and drift ashore, and whale temples can be found along the length of the coast from Da Nang to Saigon and beyond. However, whale worship in Vietnam is under severe threat from the intensification of development initiatives in coastal regions since the post-1986 Đổi Mới reforms, not least as a consequence of a booming tourism sector, resulting in the resettlement of coastal fishing villages. Parnwell observes how this has affected fishers' intimate relationship with the sea, which has traditionally been mediated via the whale genie, a spirit that is venerated in much the same way as ancestors are worshiped by land-based Vietnamese. Based on visits to 20 whale temples and interviews in 15 fishing communities in the vicinity of Da Nang and Hoi An, Parnwell discusses how rapid tourism development and occupational shifts are affecting whale veneration as a traditional cultural practice. The analysis is set against a backdrop of claims made by the Vietnamese government of protecting 'intangible cultural heritage' since becoming a signatory to the UNESCO Convention on the Safeguarding of the Intangible Cultural Heritage in 2005.

In Chapter 6 'The march of the monks: on the political repertoire of Buddhist monks in Myanmar and Sri Lanka', Michael Hertzberg explores the various ways in which Buddhist monks have chosen to involve themselves politically in different circumstances. By scrutinizing recent episodes and events in Myanmar and Sri Lanka, he considers whether we can find a common ethos among Buddhist monks concerning political engagement. Hertzberg notes the contexts are diverse. Monks in Myanmar have opposed and participated in acts of de-legitimisation of the military regime. In contrast, monks in Sri Lanka chose in 2004 to form their own political party, the Jathika Hela Urumaya, to seek a solution to the longstanding civil war against the Tamil Tigers. The chapter particularly focuses not so much on why these monks choose involvement as to how. Thus, his approach is to discern the dynamics around monkhood in relation to political involvement, and how the Buddhist robe may hinder but also enhance various forms of political actions. By asserting that there is a unique political space for the Buddhist clergy, Hertzberg hopes to establish the political repertoire that is available for monks to act upon.

Moving back again to China, in Chapter 7 'The second cultural revolution?: individualization and the emergence of psychology in late modern China', Simen Øyen and Ole Jacob Madsen overview the academic literature on individualization in contemporary China, and examine shifts in the ideals of the Chinese subjectivity. They argue how over the last couple of decades, China has evolved from an authoritarian state into a late modernised capitalist economy. This development has also set off deep structural changes regarding the ideals and strains on selfhood, comparable to previously observed cultural revolutions in the West in the wake of the sexual uprising and birth of countercultures throughout the 1960s. Of particular interest is the subjective turn which has led to the eruption of a booming psychological market targeted both to the promises and pitfalls of the newly exposed Chinese 'enterprise self'. The authors contend that results of analysis of therapeutic culture in China reveal that Chinese style individualization is both similar and different to previous manifestations in Western countries.

In Chapter 8 'National policy, sub-national trajectories: development of local models in China's health reform process', Lewis Husein examines how the management of many areas of China's social policy, including health care, is highly decentralised. The chapter studies the New Cooperative Medical Scheme (NCMS), introduced in the early 2000s, and how China's size necessitated a local adaptation of the scheme, and moreover how the grassroots 'experience' developed since the reform process of the late 1970s contributed to its subsequent reform and a progressive refinement process. Husein notes that there is increasing recognition that a dichotomous top-down versus bottom-up analytical framework is limited in understanding Chinese policy processes, and that experimentation forms a core part of the Chinese policy development repertoire and of systemic resilience. To date, though, there have been few detailed analyses of the local processes underlying systemic experimentation and innovation. This chapter provides an analysis of one county's development of an NCMS oversight mechanism in relation to reform at larger scales in Chinese government, and the extent to which this mechanism is having, or could have, supra-local applicability and/or impact.

A further analysis of local level development issues in China is offered by Adam Tyson in Chapter 9 'Horizon of riches: development and difference in Xinjiang, China'. The province of Xinjiang lies to the far west of the country. The chapter hence looks beyond China's urban east coast Special Economic Zones to a conflict-prone and resource rich frontier of Chinese society. Tyson argues that Beijing cannot simply deliver development to peripheral prefectures in the hope of resolving long-standing social conflicts and diversity dilemmas. Uyghurs and other minorities in Xinjiang have not hitherto fully enjoyed the residual benefits of the government's investment and development plans. Minority groups, rather, have their own ideas about human betterment, religious devotion, and the appropriate pace of development. This chapter contends that such incongruities cast doubt on the efficacy of what is termed Beijing's 'horizon of riches' strategy. Chinese academic perspectives are consulted in this analysis, made accessible through Tsinghua Tongfang's China National Knowledge Infrastructure database, as well as Ori Probe's China-Asia On Demand document delivery service.

Alternative insights into China's contemporary development are offered in Chapter 10 'Mapping the ethical terrain of Chinese aquaculture: exploring sustainable aquaculture trade in a global context' by Scott Bremer *et al.* They note how the progressive opening of China to international trade, and the attendant rapid modernisation and industrialisation of the country, has had implications for Chinese conceptions of the good life that should be recognised in giving effect to 'ethical' and 'sustainable' trade, which fulfils the hopes and aspirations of Chinese producers. This chapter maps the changing ethical terrain in China with reference to the important sector of aquaculture, and the 'value chain' representing the trade of aquaculture products from China to Europe. After introducing the research project 'Sustaining Ethical Aquaculture Trade', the paper describes empirical research undertaken in China and the key lessons learned. Of central importance is the need to depart from asymmetrical trade-certification schemes, which emphasise the interests of consumers in concepts of 'ethical trade,' to encompass the interests of producers in China also.

The geographic focus then shifts in South Asia in Chapter 11 'The growth trajectory of Indian multinational enterprises' by Surender Munjal, Peter J. Buckley, Peter Enderwick, and Nicolas Forsans. They chart the development trajectories of Indian multinational enterprises (MNEs) with links to the liberalisation of the Indian economy, arguing there has been a positive correlation between MNE development and the nation's liberalisation policies over recent years. However, they also contend that initial protectionist policies implemented before liberalisation helped foster these companies in their formative stages of internationalisation by providing a large sheltered home market. Later, the liberalisation of the national economy allowed Indian MNEs to establish linkages with foreign firms and subsequently accelerate their growth trajectories. The chapter also discusses the extent to which the development strategies of Indian MNEs were based on the acquisition of foreign firms. A multiple case study approach is adopted with three illustrative cases providing comprehensive and longitudinal perspectives on the subject matter.

Finally, in Chapter 12 'Renewable energy and emerging low carbon development strategies in East Asia' Christopher Dent presents a study on the rapid development of East Asia's renewable energy (RE) sector over the last decade or so. He makes the case for how renewables are forming an integral part of the region's new industrial policies and 'new developmentalism', which are founded on new configured forms of state capacity shaped in response to various challenges, primarily climate change, energy security, globalisation and global neo-liberalism. According to Dent, studying the recent progress of East Asia's RE sector provides useful insights into new emerging features of East Asia's political economy, especially in relation to the region's prospects for transition towards low carbon development. The chapter considers how and why have different approaches to RE policy emerged in East Asia, to what extent is the promotion and expansion of East Asia's RE sector part of a new industrial policy paradigm and 'new developmentalism', and what can the study of East Asian policies on promoting renewable energy tell us about the region's broader approach on sustainable development. Although the promotion of renewable energy has been a fundamental part of East Asia's recent macro-development plans and new developmentalism generally, these same plans suggest that East Asian states will simultaneously continue to significantly promote high carbon and ecologically damaging industrial activities, thus undermining the low carbon credentials of East Asia's new developmentalism. The path to meaningful low carbon development will be very long and, if achieved, will take many decades. However, it is contended that by maintaining and improving their various forms of state capacity over time, East Asian states will be well positioned to sustain the significant growth of their renewable energy sectors and thereby further strengthen the low carbon development orientation of their new industrial policies, macro-development plans and strategic economic thinking.

Core themes

Each chapter sees or views development from its own unique perspective, and this book celebrates such diversity. There are, however, a number of core themes that connect chapters together in different ways, and these are discussed below.

Networks, culture and social relations

Network formation and development is a feature of many of the chapters from both historic and contemporary perspectives, and we can see strong continuities in network evolution over time when charting various paths of 'development' in Asia. This process has in turn been shaped and determined very strongly by culture and social relations, both local within the region and intermediated with international and transnational factors. Chapters 2 (Brautaset) and 4 (Skeie) examined for example the transnational links between China, Norway and the wider world that had been forged by merchants and missionaries over time. Local culture and social relations played a key role in shaping the nature of the development of these transnational relationships. As Chapters 5 (Parnwell), 6 (Hertzberg), 8 (Husein) and 9 (Tyson) show, they were also fundamental determinants in how development was perceived and understood, or in other words how 'development' was socially constructed. There is no universal definition or determination of development, as the great diversity of the Asian experience clearly reveals.

Learning, knowledge and innovation

Following on closely from the first core theme, most Asian societies have afforded significantly high socio-cultural value to learning and knowledge as a means to achieving personal and societal improvement, or 'development' *per se*. This is evident for example in Asian education cultures and systems, and the importance to Asian organisations of acquiring learning and knowledge from both domestic and foreign sources, or through transnational processes. The extent to which Asian societies and organisations have developed the same level of innovative capacities as their Western counterparts is a more contentious matter, but this depends on how we define 'innovation'. Moreover, it is evident that even Asia's developing countries are fast improving their propensity for innovation, which in turn will shape the future path of the region's development. This was discussed for example in Chapter 12 (Dent) in the context of renewable energy systems and technology, and their contribution to Asian efforts on low carbon development. Chapter 11 (Munjal *et al.*) meanwhile examined the importance of learning, knowledge and innovation in the internationalisation and development of India multinational enterprise companies. Chapter 3 (Medvedeva) discussed how foreign knowledge was intermediated in the historic development of the Chinese customs service, while Chapter 7 (Øyen & Madsen) explored a similar dynamic in the contemporary development of emerging Chinese cultures of individualisation.

Modernity and post-modernity

There are a wide range of debates and discourses on modernity and post-modernity involving various disciplinary perspectives. From a political economy view, for instance, Asia's economic development over recent decades could be said to be firmly grounded in the principles of modernity or modernisation, where rapid industrialisation has delivered improvements in material-based prosperity through resource-intensive means, and the subordination of nature and environment to the needs of humankind. This of course applies to other regions, and the global contemporary practices of 'modern' economic development. Post-modern models of development include more enlightened appreciations of the relationship between humankind and environment, and the imperative of paradigm shifts towards sustainable development at multiple levels of practice and approach. There is also the ecological modernisation and late modernisation approaches to consider that do not prescribe to radical or revolutionary systemic changes to the economy or society, rather perpetual reform to existing structures. How may we view Asia's development over time in these respects, and how may also 'pre-modern' perspectives be relevant in both historic and contemporary contexts? Tensions between tradition and modernity are a recurring and widespread feature in stories of Asian development, perhaps best exemplified in Chapter 5 (Parnwell) and Chapter 7 (Øyen & Madsen). Whether Asia's efforts to establish cleaner forms of energy generation and economic development were just an extension of the region's modernisation were discussed in Chapter 12 (Dent).

Sustainability and security

This is closely related to the above theme. To what extent are sustainable models of development emerging in Asia, and what are their prospects for gaining stronger ground as the 21st century unfolds? Furthermore, what security risks (widely conceived and at various levels) have arisen over time and will arise in the future concerning the trajectories of Asian development? This relates to various sectors of security (e.g. human, politico-military-maritime, energy, environment, etc.) and may be strongly linked to issues of sustainability. Chapter 10 (Bremer *et al.*) discussed this in relation to sustainable aquacultural practices, Chapter 12 (Dent) concerning energy and Asia's development, Chapter 9 (Tyson) concerning how China is managing the sustainability and security aspects of provincial development, and Chapter 5 (Parnwell) from the perspective of the sustainability and security of local livelihoods.

Geo-scale: localism, transnationalisation and globalisation

Asia has a long history of interactions with transnational and global systems. Highly developed regional trade systems were extremely well established in Asia long before the maritime European powers arrived on the scene. While Asia's 'great diversity' may be essentially defined by multifarious distinct local cultures and identities, its development has too been intermediated by the aforementioned connections and interactions. Asia's place in today's globalising world – despite its highly asymmetric pattern of transnationalised activities and structures – is a very important one, and as Asia's significance grows in the 21st century will only become more so. Understanding the long trajectories of Asia's development in an integrating world over time is critical to understanding world futures. Chapter 2 (Brautaset) and Chapter 4 (Skeie) examined the conceptual as well as historical development of transnationalised links between China and the rest of the world, while Chapter 11 (Munjal *et al.*) explored this theme from the perspectives of Indian companies. Transnational food supply chains were the subject of discussion in Chapter 10 (Bremer *et al.*). At the same time, Chapters 3 (Medvedeva), 5 (Parnwell), 8 (Husein) and 9 (Tyson) showed how development is ultimately rooted in the local, and thereby determined by factors of local diversity however conceived.

Chapter 2: Merchants and missionaries

Connecting China, Norway and beyond

Camilla Brautaset

Globalisation and China's draconic economic growth have been buzzwords for the media and academia alike over the past decades. Questions like 'When did globalisation really begin?' and 'When and why did the great divergence emerge?' have been central to these debates. China in the late Qing Dynasty has been at the heart of both discourses. During the 'treaty century'[1] from the 1840s and until 1943, China developed into a global meeting place. A substantial part of Western research on China's interaction with the West during this period has revolved around actors from major powers such as the British Empire, and has been studied through the lens of imperialism (for instance Bickers 1999, 2004 [2003], 2011, Brunero 2009 [2006]). These works have offered invaluable contributions to the understanding of the foreign communities in the Middle Kingdom. However, actors from small nations mattered too.

Norway was one such small country. Despite being a geopolitical minnow, it played a disproportionate role in international shipping and was an integral part of protestant mission. Around 1880, Norway had the third largest merchant fleet in the world and religious movements increasingly directed their attention towards global mission fields. In the early 1890s, both merchants and missionaries set their eyes on the largest market in the world. China soon developed into a newfound stronghold for both groups and their activities were to flourish over the next decades. In 1937 the genocide in Nanjing signalled that the second Sino-Japanese War had entered into a stage of 'total war'.[2] This was also the beginning of the end of an era for the many Norwegians living and working in China at the time. As time has passed by, their stories and experiences have fallen into oblivion. Now, the time has come to tell the largely forgotten history of a period when China and Norway were intimately entwined by networks forged by merchants and missionaries. This is the core of what we hope to do in the project that is at the centre of this chapter; *Merchants and Missionaries. Norwegian encounters with China in a transnational perspective,*

[1] The 'treaty century' refers to the period from treaty of Nanjing in 1842 to the abolishment of all foreign extraterritoriality in 1943 (Fairbanks 2006 [1992]).

[2] 'Total war' is commonly understood as all-inclusive, fully mobilised and unrestricted warfare, and where there is no distinction between military and civilian targets.

C. Brautaset and C.M. Dent (eds.) **THE GREAT DIVERSITY**
DOI 10.3920/978-90-8686-786-8_2, © Wageningen Academic Publishers 2014

1890-1937. This project evolved from a newly established research group in 2009.[3] Some members of the group were surprised at how little work had been undertaken on Norwegian shipping in late Qing and the early republic despite contemporary publications clearly stating rapid growth and the increasing importance of the East Asian markets.[4] Others were bemoaning the lack of research from the perspective of migration, whilst historians of religion were calling for much needed research on one of the major focal points of Norwegian missions during the first half of the 20th century, namely China. To date, there has been no major comprehensive academic study of these missions.[5] Internationally, there has been a recent push towards integrating missionaries into larger narratives (e.g. Clossey 2006, Wendt 2011). However, although merchants and missionaries tend to be underlined as China's key contacts with the West from the 1840s towards the 1940s (e.g. Bayly 2004, Fenby 2008), few synthesising efforts have been made to investigate these two groups together.[6]

This is also the status of the merchants and missionaries that served as nexuses between China and Norway. These actors travelled across major geographical distances and crossed both cultural and linguistic borders. Migration literature suggests that such cross-community migrants[7] may have a much larger societal impact than their numbers would infer. The Lucassen brothers explicitly differentiate migrants settling in rural areas with limited interaction with their surrounding environment from smaller, professional groups, which '...may have had a large and lasting influence on receiving societies' (Lucassen & Lucassen 2009: 352). Merchants

[3] The research group comprises historians, historians of religion and cultural specialists. More information available at: www.uib.no/fg/internasj. In 2010, the author of this chapter launched a successful bid for funding to the Norwegian Research Council within the categories of independent project within the humanities. More information on this project available at: http://merchantsandmissionaries.no.

[4] Moreover, the interest from scholars appears to have faded over time. The two main academic contributions on Norwegian shipping in East Asia are Ytreberg (1951) and Von Mende (1971), whilst Dannevig published a popular history in 1982. However, Brautaset & Tenold (2010) have recently published a study of Norwegian vessels in Asian trades 1870-1914.

[5] However, Mikaelsson (2003), Strandenæs *et al.* (2006) and Sannes *et al.* (2008) all offer valuable studies on various aspects of Norwegian missions in China and some key individuals. In contrast to the academic research status, a vast number of popular works have been written from within the mission's own perspective. The missions own accounts do, however, follow denominational lines and has a propensity to emphasis hero-martyr motives.

[6] Two exemptions deserve special mention here. First, *Merchants, Missionaries and Migrants* (van Kessel (ed.) 2002), offers a transnational history between the Netherlands and Ghana. Second in 1983, James Reed published *The Missionary Mind and American East Asia Policy 1911-1915*. The book offers a detailed account of how merchants and missionaries impacted American foreign policy through both actions and non-actions. Reed claims that American businesses' fascination with Japan left them more indifferent towards China, whilst the missionaries' impact on foreign policy was profound. Reed argues that this impact was unintended – but disastrous through 'selective recognition and wilful misrepresentation' (Reed 1983: 198-200).

[7] Cross-community migration is understood as migratory movements that are crossing linguistic and cultural boundaries (Manning 2005, 2006, Lucassen & Lucassen 2009).

and missionaries certainly fall within the latter definition. They represented small, migratory groups with high impact potential on the societies of destination as well as departure. Yet, few – if any – existing studies apply migration theories or perspectives when analysing these groups. Applying theoretical insight perspectives from migration literature – and especially from the perspective of cross-community migration – may offer what may be perceived as a missing link in addressing the mobility of merchants as well as missionaries. For instance, missionary literature tends to emphasise that the missionaries' choice to go abroad stemmed from personal decisions and callings; however, their timing coincided with the great migration movements from the Scandinavian countries. Moreover, the Norwegian historian Francis Sejersted has described this period as an interlude of economic development, where geographic expansion rather than changing technological regime was the predominant driver for economic development (Sejersted 1993). Indeed, as a unified research universe, merchants and missionaries hold the potential to enhance our understanding of the interface between such economic and cultural processes. In this project, we aspire to do so by focussing on networks and mobility and argue that a transnational approach with emphasis on human agency enables us to capture globalisation processes from below. Before outlining further details of the project, however, we will start by looking at how the project aligns with current trends within historical research in general, and foreign-Chinese relations in particular.[8]

The big picture: globalisation, history and global history

The merchants and missionaries project addresses a period when China was an integral part of global exchange and interaction. During the late Qing Dynasty and the early republic, China witnessed an influx of foreign direct investments (FDI), extraterritoriality as well as rapid migratory flows of merchants and missionaries.[9] Still, these agents would probably not apply the term globalisation in order to describe their experiences. Just as history is a science of hindsight, globalisation was a term that emerged much later than the events it often describes. The term globalisation was not coined until the 1940s and only became a part of mainstream jargon from the 1980s onwards.[10] Hence its application to historical events should be treated with care.

[8] The project contains three sub-projects on merchants, missionaries and customs officials. The main focus in this chapter will be on merchants, whilst Medvedeva's and Hestad Skeie's contributions in this book offer further information on the customs service and the missionary enterprise respectively.

[9] For a contemporary overview of FDI in China, please see Remer 1933.

[10] Scholte offers a valuable run-through of the conceptual history of globalisation (Scholte 2005: 16, 50-52).

In an informative article assessing the historiography of globalisation studies, the sociologist Luke Martell addresses what he refers to as 'the three waves of globalization theory' (Martell 2007).[11] According to him, these waves have left at least three competing perceptions. The globalist school emphasises a quantitative approach and an economic perspective on globalisation. Here, globalisation is perceived as a phenomenon of very recent history and a source of homogenisation as well as the loss of congruency of state and territory. The sceptical school, on the other hand, questions the very existence of globalisation. Sceptics argue that internationalisation is a more applicable description of the historical processes that globalisation discourse revolves around. Internationalisation, they argue, is observed throughout the history of human civilisation, with experiences from modern history suggesting stronger rather than weaker nation states. The transformationalist school, however, has aspired to further develop the original views of the globalists by adjusting the theories in order to account for the criticism of the sceptics.[12] In short, the transformationalist concludes that globalisation is a real phenomenon with causal effect on economics, politics and culture – but where the experience of globalisation is heterogeneous in time and space (Martell 2007). According to this view, globalisation is not new as such, but the scale, intensity and ramifications of globalisation today is unprecedented in history.

Though scholars disagree on what globalisation is exactly and how it has unfolded in history, they unanimously agree on the importance of furthering our knowledge in this field. Some take this a step further in their programmatic declarations:

> Globalization(s) and, thus, global history, as a reality and not just as a concept, are now something every person has to deal with in one way or another, in their everyday life as well as in their comprehension of the world and the way that they choose to act in it. Therefore, globality comes with globalization(s) and global history, again as both a concept and as a reality (Gills & Thompson 2006: 4).

A key issue in the debate on globalisation and global history is the question of territory and political autonomy on a state level. Historians and the sceptics in the globalisation debate have argued that the combination of industrialisation, increased market integration in the late 19th century, combined with imperialism and the growth of nation states, challenges the perception that globalisation reduces national sovereignty. However, the relevant literature tends to ignore that there were two very different scenarios unravelling at the same time. Whilst nation states in the

[11] In a more recent work, Martell introduces a fourth category in terms of 'post-modern' approaches to globalisation (Martell 2010).

[12] Martell also offers a very useful tabular overview of the three schools he identifies (Martell 2007: 177).

West were being fortified, empires of the East were falling. This was the case in the Qing Empire in 1911, the Russian Empire in 1917 and later the Ottoman Empire in 1922. A similar argument could be put forward for religion. Religion in general and missionary movements in particular, are either left out or portrayed as a sidekick in mainstream narratives of globalisation in the late 19th and early 20th centuries. Osterhammel and Petersson offer us an example of the latter. They emphasise the failure of Christianity in becoming a dominant world religion despite the efforts of missionaries and European hegemony.[13] However, the authors seem to forget the success of the missionary movements in aiding '...concentration and increasing importance of worldwide integration' – which is their definition of globalisation (Osterhammel & Petersson 2005: 26). Peter Pels, on the other hand, argues that: 'Missionaries have been some of the most pervasive, powerful and persistent protagonists in the long story of globalization' (Pels 2009: 716). The complexities are striking. Political unification and devolution, secularisation and religious mobilisation appear to have been happening in parallel. The discussion above suggests that if one applies the concept of globalisation in historical research, there are several good reasons to opt for a definition that is less static and programmatic in nature than some of the definitions found amongst globalists and transformationalists. This author's understanding of globalisation combines definitions put forward by leading historians within this field: 'The continuous processes of connections, transactions and interactions between larger geographic areas' (based on Osterhammel & Petersson 2005: 26, Grafe 2010: 253).

There is an intimate link between the globalisation discourse and the genre of global history, but there are also distinct differences. In her inspiring book *What is Global History?*, Pamela Crossley argues that global history is more about methods and concepts than facts (Crossley 2008). As Crossley points out, the majority of authors of global histories are not historians by training. According to her, the mandate of global history is not to provide new empirical arguments as such, but rather to offer perspectives on the basis of existing empirical research:

> In this basic foundation of training and the commitment to generate original knowledge, the historian and the writer of global history have little in common. There is nowhere to go and no method to use for researching global history, since there is no global context for the generation of evidence (Crossley 2008: 105).

[13] Failure here appears to be defined as the lack of political capital and remaining a religious minority in large parts of the World. As for Christian missionary efforts in Asia, the same authors argue that they'...bore no fruit' (Osterhammel & Petersson 2005: 75).

In un-distilled form, global history is history without a centre. In consequence, historical sources are dispersed and in principle infinite in their abundance. Hence, in this perspective, most historians would be handmaidens to global history in terms of supplying empirical studies, historiographical reflections and critical discourse. Within the discipline of history, however, the above-mentioned rise of global history and global perspectives have also been a reaction to what has been perceived as a Eurocentric hegemony – especially within the sub-discipline of economic history. In traditional Eurocentric history writing, China and Asia (along with essentially the rest of the World) have either been overlooked or used as an antithesis to Western greatness (e.g. Brenner 1976, Jones 2004 [1981], Landes 1999 [1998]).

Over the past decades, the Eurocentric view within world history has become increasingly challenged. In the post-war era, much attention was paid towards explaining the 'European miracle'. Today, world history is being rewritten in order to account for Asian economic development in general and China's development in particular.[14] As a result, the debate on the great divergence between the West and the rest has moved on from intangible determinants such as values and culture, to also include tangible factors such as luck, colonies, and the existence and location of fossil fuels. The so-called California school[15] has been particularly industrious in promoting research suggesting how, for instance, China enjoyed markets on par if not more sophisticated than their Western counterparts pre 19th century (Frank 1998, Goldstone 1993, 2008, Marks 1998 [1997], 2002, Pomeranz 2000, 2001, Wong 2000).

The Eurocentric picture of China in terms of politics and international affairs before the Opium Wars is one of an inward-looking empire exerting absolute power through a centralised bureaucracy. This has been described through the much used and continuously re-interpreted Western concept of Oriental despotism.[16] Few, if any, of these stereotypes hold true against compelling historical evidence that has been brought forward by historians and China specialists over the past decades. Now, the understanding of China as stagnant, despotic, backward and operating in splendid

[14] Presentism continues to be a hot potato amongst historians. Some scholars have argued that history's epistemological status is threatened by the demands of rewriting history in order to account for the present (Dirlik & Bahl 2000: 3).

[15] For an excellent recent overview and assessment of various schools on China's economic development, see for instance Brandt *et al.* (2013).

[16] It is impossible to do justice to Oriental despotism, its historicity, numerous applications and its profound impact on European presentations and representations of Asia in just a few lines. However, the concept can be traced back to Antiquity. In modern times, it is particularly the works by Hegel, Marx, Weber – and later Wittfogel – that have defined this debate. For an engaging account of the historical development of the concept see for instance Minuti (2012).

isolation pre-1830 is in itself becoming a subject of the past. The research on China and its relations with the rest of the world has also seen the emergence of a new generation of historians that not only challenge Eurocentric stereotype perceptions of China, but also call for a new, revisionist history of China to be narrated. This direction emphasises how late Qing China changed as result of forces within as well as those imposed from outside; it focuses on cultural transformations and hybrid identities – and it calls for moving beyond and below political history *per se* in order to better understand Chinese interactions with the wider world. The eminent historian and China specialist Odd Arne Westad argues that the conventional histories of China have had a propensity of focusing on the state apparatus, wars, diplomacy and top-down processes, and thereby failing to capture the processes that shaped most people's experiences with foreign and international relations. Specifically, Westad calls for a history that '...gives equal treatment to missionaries and diplomats, businessmen and revolutionaries, workers and bosses' (Westad 2012: 13). Hopefully, the history of Norwegian merchants and missionaries will be a small, but vital contribution towards this direction.

The project: merchants, missionaries and entwined histories

We do not know exactly when the first contact between China and Norway occurred. What we do know, however, is that Norwegian sailors were onboard Dutch and Danish vessels anchoring on the shores of China in the 17th century (Myrstad 2009, Von Mende 1971).[17] The contact remained sporadic until well into the 19th century after which everything changed. Norway's merchant marine was growing rapidly from the 1830s onwards through systematic re-investments of earnings from Norwegian and Swedish foreign trade, in particular timber (Brautaset 2002, 2011, Brautaset & Tenold 2008). The repeal of the navigation acts from 1850 signalled the opening up of the international freight market. From then on, there was no looking back for Norwegian ship-owners. Cross-country (i.e. freight between foreign ports) shipping became the most important market segment, and Norwegian vessels chased freights in ports all over the globe. Norwegian ship-owners were rarely pioneers in exploring new markets, but were literally happy to go with the flow in the pursuit of business opportunities in a brave new world (Brautaset & Grafe 2006).

This was also the case of East Asia and China. Norway's access to the Chinese market came in the wake of the intervention by major powers. The defeat in the First Opium

[17] The Dutch East India Company [Vereenigde Oost-Indische Compagnie], was particularly instrumental in this respect – its labour force was as global as its business. Norwegian sailors were onboard vessels heading for India, Batavia [Indonesia], Japan and China. Interestingly, the three latter destinations were to be the cornerstone of Norwegian shipping in Asia two centuries later.

War (1839-42) forced China to sign treaties with major Western powers. The treaties opened five ports and allowed for extraterritoriality[18] for the treaty nations.[19] As highlighted in the introduction, this was the start to the treaty century of Chinese history. Unknown to most international scholars, Sweden-Norway[20] also obtained a treaty of its own in 1847.[21] In the early 1850s, Sweden-Norway established a general consulate in Canton, Guangzhou, and a vice-consulate in Shanghai (Myrstad 2009).[22] The so-called Arrow War (1858-60) resulted in a new set of treaties being signed between China and the West. The new peace settlement of the Tianjin treaties opened another eleven ports to foreign merchants in addition to furthering the privileges already given to Western merchants and missionaries.[23] This event also resulted in the setup of an entirely new Chinese agency in 1861; the Zongli Yamen, whose mandate was to deal with foreign businesses operating in China (Spence 1999 [1990]).[24] Hence, foreign businesses were not only given a competitive advantage with regards to Chinese enterprises through the duty system, but they were also given an absolute institutional advantage too.[25] Still, institutional pre-conditions were necessary, but not sufficient to trigger an immediate eastward flow of Norwegian merchants and missionaries. Although the opening of the Suez Canal in 1869 had a profound impact on connecting Europe with Asia, the Norwegian presence was not

[18] Extraterritoriality implied that citizens of treaty nations were exempted from Chinese law. Extraterritoriality as such was not a new phenomenon in China, however, the scale and scope of the extraterritoriality was unprecedented and remains controversial in China today.

[19] The American treaty of 1843 came just a year after the British treaty and paved way for American protestant mission. Spence highlights how article 17 symbolised this through giving the Americans access to build cemeteries, churches and hospitals in the treaty ports (Spence 1999 [1990]: 163).

[20] Sweden-Norway was in a loose, personal union 1814-1905. The Foreign Service was the only joint administrative body of the two countries.

[21] The Swedish-Norwegian treaty of 1847 has been portrayed as the only so-called 'unequal treaties' achieved without gunboat diplomacy. The treaty gave Sweden-Norway extraterritorial jurisdiction and immunity of its citizens to Chinese law, lower tariffs on trade and commerce, the right to purchase land, to build churches, and the right to learn Chinese in these ports. A conclusive part of all these treaties was the removal of tariffs on vessels registered in the treaty countries. Similar concessions were not given to Chinese vessels. Recently the historian Pär Cassel has offered new research on the judicial status of the treaty and pointed out that the Qing empire never formally ratified the treaty. However, Cassel argues convincingly how the treaty still governed the political relationship between the involved parties until new treaties came in place in 1908 (Cassel 2010).

[22] The Consulate General was already moved to Shanghai in 1863.

[23] During the 1860s several efforts were put in place in order to prepare a revised version of the treaty of 1858. However, the British House of Commons turned down the revised version of the treaty in 1870 (Spence 1999 [1990]).

[24] It would take another four decades, however, before an own Chinese corporate law was in place. In 1904 the recently established Ministry of Commerce [Shangbu] issued the empire's first company law, the Gongsilü. The corporate legal system went through several reforms prior to 1937, most notably in 1914, 1929/31 – including the introduction of China's first Maritime Law in 1929.

[25] Several scholars, and maybe most notably the historian and sinologist William Kirby has emphasised legal, institutional factors in explaining why foreign businesses blossomed in China, whilst similar large, Chinese corporations were notably absent (Kirby 1995: 43-63, especially 44-48). This is often referred to as the 'Kirby Paradox'.

significant until the 1890s when Norwegian merchants and missionaries channelled their efforts towards booming Chinese markets. Several Norwegian firms were set up in China, primarily as subsidiaries of companies registered in Norway. The most prominent companies were the trading company Oversøiske Compagnie (The Oversea Export & Import Co. Ltd) established in 1898 and Aktieselskabet Union Bryggeri AS (Scandinavian Brewery Co Ltd). [26]

The major commercial activity was, however, not in the trade of goods but in services. Along the Chinese coast, Norwegian vessels were increasingly vying for freights in the bustling ports. From the 1890s onwards, ship-owners and agents made their presence permanent through opening offices in economic centres and boomtowns such as Shanghai and Hong Kong. In the beginning of the 20th century, Norwegian ship owners had developed a stronghold in certain market segments, such as the rice trade between Indo-China and Hong Kong, the coal trade between China and Japan and parts of the Chinese coastal trade. Contrary to the business model applied in the Atlantic trades, the key components in the Asian and Chinese market was cabotage[27], charter parties and steam. Traditionally, Norwegian ship owners had specialised in the spot market of shipping through the tramp trade. In East Asia, however, most Norwegian vessels were deployed through time charter (Brautaset & Tenold 2010). This implied that whilst the ship owners were still responsible for the management of the vessel, the operational control – including securing freight – was left to the charterer.[28]

The development of Norwegian shipping in China mirrors its transnational business mode by suggesting dependency upon primary and secondary economic swings

[26] Oversøiske Compagnie was listed in Kristiania [Oslo], whilst operations were carried out through its Shanghai office and hence corresponding with definitions of a freestanding company (Wilkins & Schrøter (eds.) 1998, Jones 2000). There is currently no history of Oversøisk Compagnie. Einar Bjørnson, Simon Arnold and Oluf Thoresen, who all had lived in China for some time, founded the company during the first stock market boom in Norway in the late 1890s. Twenty years later, it was one of the largest industrial companies in Norway in terms of paid up capital and had branches in 25 of the major ports around the world. The company went spectacularly bankrupt, however, in the 1920s after speculation in sugar in Java went horribly wrong.

[27] In the context here, cabotage refers to regional trade encompassing China, Japan, Russian Asia as well as what the Chinese refers to as the region of Nanyang (Becker 2010: 255).

[28] The Norwegian ships in the China trade had typically Norwegian captains and officers, a Chinese crew – and often a Chinese comprador to assist in translation and when at port. The duration of the charter parties in the East Asia trades was a maximum of two years and typically one year. In contrast, the charter parties in the fruit trade in America usually lasted for three to four years (Schreiner 1963: 28f).

in China as well as Europe.[29] Figure 2.1 contains calculations based on Liang-Lin Hsiao's monumental work on Chinese foreign trade statistics (Hsiao 1974). The figure offers a simple comparison of tonnage arriving on Norwegian keels compared to the corresponding figures for overall shipping arriving at Chinese ports.[30] There are at least three distinct sub-periods. First, the period up to 1912 can be described as Expansion (ca.1890-1912). The advance in the China trade had both a technological and a market explanation. Inspired by the consular general in Shanghai, Norwegian ship owners were increasingly deploying purpose-built steam vessels for the China trade from 1893 onwards. They become an instant success and a cash cow for their owners as their launch coincided with high demand and peaking freight rates during the first Sino-Japanese War. Inspired by the profits made from war, several Norwegian shipping companies were established not long thereafter. These firms were primarily offering brokerage and agency services, such as for instance Oluf Thoresen & Co.[31] (1898) and Haakon Wallem & Co. (1903), which were both founded and based in Shanghai.[32]

The Russo-Japanese War 1904-1905 marked the peak of Norwegian shipping in China with over 1,500 registered arrivals in Chinese ports. As can be seen from Figure 2.1, the phase of rapid expansion was subsequently followed by a period of Contraction (1911-1920). The comedown was a result of a conjuncture of factors and events such as the loss of prime markets such as the important coal trade with Japan, the burst of the Shanghai rubber boom in 1910 and the opportunity costs of not having the vessels deployed in the European trade. Ship owners had already started pulling out of the trade in the northern ports from 1906. Some vessels returned to Europe, others became temporarily deployed in trade in southern Chinese ports, Indochina and East India (Ytreberg 1951). Some vessels also sought trade in the East-Russian markets, but like Japan – Russia also soon imposed preferential treatment to domestic vessels in the cabotage trade.

[29] The economic historian Jan Tore Klovland has found four clearly marked peak years in his excellent study of the international freight market before 1914; 1873, 1889, 1900 and 1912 – which all mirror business cycle peaks (Klovland 2002). The cycle peaks – which are all heavily based on Europe and the Atlantic – of 1900 and 1912 are especially interesting as they co-inside with major negative economic and political shocks in China, i.e. the Boxer Uprising and the Revolution of 1911 and the establishment of the Republic in 1912.

[30] It is, however, important to stress that when dealing with the overall picture, the market share of Norwegian shipping never exceeded 8% (peaking in 1905 with 7.8%). Hence, in order to compare the relative development over time, the volume development is displayed as Laspeyre indices with 1910 as the base year.

[31] It was initially a subsidiary of Oversøisk Compagnie.

[32] The first Norwegian with a ship-owning company in China was Jens Andreas Danielsen, who set up a company in Hong Kong in 1884. Danielsen was followed by Anders Olsen, who set up the shipping company Olsen & Co in Shanghai in 1894 (Seeberg & Filseth 2000: 51, 57).

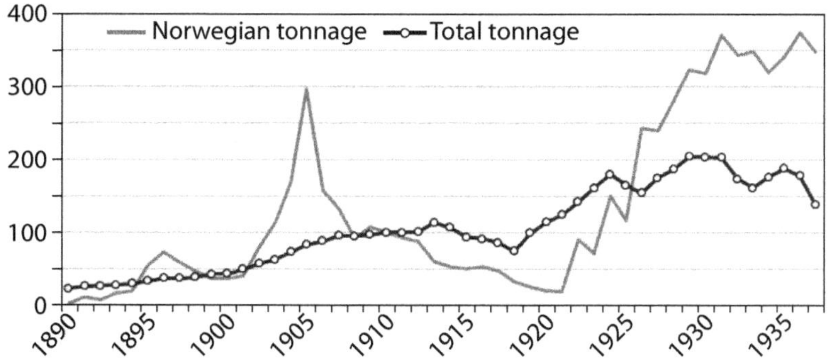

Figure 2.1. Norwegian shipping in China ca. 1890-1940 (1910=100).

However, a period of Resurrection (1921-1939) was to follow as Norwegian vessels again called in Chinese ports. Indeed, the most dramatic expansion in terms of volume came in the early 1920s, this time in terms of multi-fleet companies, specialising in certain niches – especially the passenger trade. A prime example of this expansion was one of Norway´s largest shipping company, the Drammen, based in Bruusgaard Kiøsterud and which in 1922 started their own passenger line – the Chino Siam line (Helle 2012). Bruusgaard Kiøsterud was not the only comeback kid in the China trade in the 1920s. Several other agents, brokers and ship owners such as Haakon Wallem were also returning to Chinese shores in the wake of economic austerity and imploding shipping markets in Europe.

The service industries in China served both as a contingency plan and as a target in its own right for other groups of actors too. Military trained young men and seamen migrated to the Middle Kingdom in a quest for a better life and to pursue a career in the Chinese maritime customs. Between 1860 and the 1930s, more than 300 Norwegians were engaged in this service. As has been outlined elsewhere in the literature, the hierarchical structure of the organisation offered very different career paths and rewards to its staff (Brunero 2009 [2006]).[33] Although most Western literature tends to emphasise the strong distinction between indoor and outdoor staff, another schism was between Chinese and foreign staff. It was not until the stewardship of Sir Fredrik Maze during the nationalist period (1929-1937) that Chinese staff were given the opportunity to serve as commissioners (Chang

[33] The Chinese customs have been attracting substantial academic attention recently. At Bristol University, Robert Bickers and his team have been working on customs history for more than a decade and generously share their resources with the wider community of scholars: see for instance http://www.bristol.ac.uk/history/customs. Moreover, there is a vibrant academic community studying the customs at the University of Xiamen. Harvard University has recently been undertaking a major digitalisation project of key sources of the customs service in cooperation with Wu Songdi at Fudan University. Searches are available at http://hollis.harvard.edu.

2013). As for Norwegians in the customs service, they were recruited from opposite ends of the social spectrum. Outdoor positions such as low-ranking watchers and tidewaiters tended to be filled by seamen. Spot checks of staff lists indicate that Norwegians were working as outdoor staff in ports all along the China coast.[34] The average time in service seems to have been relatively short compared to indoor staff. The reasons for leaving the service varied from resignations, dismissals, discharges, death in service or simply individuals disappearing. The findings may also suggest that the customs service was a safety valve for foreign sailors waiting for employment on vessels. Norwegians working as indoor staff were, however, recruited through entirely different channels. This was a career opportunity for an emerging and increasingly self-aware middle class, and where cultural capital played a decisive role. They were typically recruited after graduating from the naval academy or as younger members of the urban elite, and also tended to stay longer in the service.[35]

The missionaries, on the other hand, were recruited from a broader and more heterogeneous social spectrum. However, as Hestad Skeie emphasises in her chapter in this book, female missionary workers in the main Norwegian mission organisations were initially recruited from the upper social strata – though this was to change over time.[36] In China, the Norwegian merchants and missionaries settled far apart. Whilst merchants, sailors, captains and customs officials nestled along the coast and the main waterways, the missionaries found their habitat in the Yellow River valley in the Central Plain region. From a modest start in the early 1890s, this – allegedly the home to Chinese civilization – was to become a major missionary focal point to the Norwegian missions.[37] The missionaries found their way to China through transnational and non-denominational missionary organisations such as the China inland mission[38], Scandinavian-American organisations as well as

[34] Annual publications with detailed information on staff can be found at the archives of the SOAS library in London. Digitalised versions of these publications for selected years can be found here: http://pds.lib.harvard.edu/pds/view/44874010.

[35] For an extensive report on Norwegians in this service, please see Hopstock (1938). PhD research fellow, Olga Medvedeva at the University of Bergen is currently undertaking the first systematic study on this group. Medeveva´s preliminary results indicate that Hopstock´s report suffers from inaccuracies and lacunas. She is developing a prosopographical database of all Norwegians in the service from ca. 1860-1940.

[36] To women from the working classes, foreign mission organisations such as the China Inland Mission could offer greater possibilities of social mobility during the 1880s and 1890s.

[37] The first Norwegian mission organisation, Det Norske Misjonsselskab (DNM), was founded in 1842. Its first missionary, Hans Paludan Smith Schreuder (1817-1882), travelled from Cape Town to Hong Kong during the winter 1847/1848 in the hope of taking up missionary work in China. However, he soon returned to South-Africa, disheartened by his own possibilities as a missionary in China due to his blond and fair physical appearance (Myklebust 1980: 56f.) Still, he urged DNM to make China the primary target of their mission. The next missionary from DNM was not, however, to appear in China until over 50 years later.

[38] Alvyn Austin claims that the image of China portrayed by founding father of China Inland Mission, Hudson Taylor, came to dominate Western perception of China in the late 19th century and beyond (Austin 2007: 3).

Norwegian missionary organisations.[39] The China-fervour in Norway during the 1890s led to five Norwegian missionary organisations taking up work in the Yellow River valley – in particular in the provinces of Hunan and Hubei – between 1889 and 1902.[40] A similar number of US-Norwegian missionary organisations were set up during the same time frame, targeting and settling in the same regions of China (Ekeland *et al.* 1919). The religious differences between the organisations, together with the general development of Norwegian Christianity towards the turn of the 20th century, suggest an increasing organisational differentiation along denominational lines. However, the missionaries' educational background, and the fact that some missionaries worked in more than one missionary society at different points in their careers, suggests a more dynamic and transgressional flow of ideas and people than is generally acknowledged.[41] In his masterful study on American protestant missions 1880-1930, Ian Tyrrell underlines how the mission movement was explicitly transnational in its outlook and ideology (Tyrrell 2010).[42]

This flow of capital, vessels, merchants and missionaries appears to have been one-way traffic. Norway did not witness a similar influx from China. However, Chinese dignitaries were amongst the first to visit the newly independent Norway.[43] The arrival of the Chinese Foreign Mission in Oslo during the spring of 1906 was hailed by media, businessmen and government officials alike.[44] The official mandate of the Chinese delegation was to study foreign political systems, in particular constitutional set-ups. However, as the reports from both Norwegian and Chinese sides show, this was much more than just a political encounter. The local hosts mobilised royalties, politicians, preeminent bureaucrats, leading business figures as well as media.[45] In particular, the Norwegians mustered large resources in promoting their own business in China and no business more so than shipping. In contrast, the missionary organisations were not represented at any of the events, let alone found around the

[39] The American Carl Remer claimed that missionary infrastructure accounted for more than half of Norwegian FDI in China (Remer 1933: 663).

[40] For a detailed account Norwegian mission in China please see Hestad Skeie´s chapter in this book.

[41] For instance, evangelical texts were intended to speak to all Protestants and would-be Protestants across the world. Missionaries mobilised major efforts in translating evangelical texts into local languages, an enterprise symptomatic of their recurrent movements between transnational, national and local contexts.

[42] Tyrrell argues convincingly how transnational networks contributed towards shaping an American non-territorial empire and cultural hegemony (Tyrrell 2010).

[43] Norway gained her independence from Sweden during the summer of 1905.

[44] The mission was the result of several Imperial edicts issued during the summer of 1905. A total of five special commissioners were selected with the mandate of studying political mandates abroad. The mission to Norway was lead by Duan Fang and Dai Hongci, both leading and reform friendly politicians in the late Qing dynasty.

[45] Fredrik Schjøth, a former commissioner to the Chinese maritime customs, was amongst four especially selected official attachés to assist the Chinese delegates. The other attachés were major general Johan Lauritz Bull, Georg Von Tangen from the Foreign Office and Norwegian vice consul in Hong Kong, Jørgen Jacob Eitzen.

dinner table.[46] There may be many reasons behind the non-show of missionaries. On one hand, the scale of the Norwegian missions in China was still relatively limited. On the other hand, however, the Chinese visit happened in the wake of the Boxer uprising and missionary activities were still highly controversial within the Qing Dynasty. Yet another explanation may be that ship owners and customs officials were able to draw upon their political and cultural capital through personal networks, whilst missionary organisations did not have the same opportunities. There is of course also the possibility that there were conflicting interests between business and religion – sharing a Norwegian background may have given rise to conflict as well as co-operation.

Connectivity

This leads us to the question of the extent of personal or institutional networks between the merchants and missionaries. These are questions that have yet to be answered. However, we do know this: these groups shared a simultaneous turn towards China. And, as suggested in the introduction, their movement coincided with the great migration wave from the Scandinavian countries in the late 19th and early 20th centuries and Sejersted's description of a spatial expansion for Norwegian businesses. Moreover, these actors shared contexts – at home and on their journeys to China, but to a lesser degree within China itself. Though personal links between these groups are interesting in their own right, they are not the main objective here. Rather, our ambition is to offer a synthesising study of merchants and missionaries through a coherent meta-framework of analysis. This framework consists of two primary components; transnational history and network analysis. The transnational approach is inspired by Yves-Pierre Saunier's claim that transnational history enables us to capture globalisation processes from below (Saunier 2009). Moreover, unlike global history in a macro perspective – transnational histories do have centres and there are places to go if one aspires to generate original knowledge [as Crossley puts it]. In the context of the merchants and missionaries project, transnational history is not understood a method as such, but rather a genre within the broader historical discipline. More specifically, it refers to historical analysis of people, phenomena, places, goods, information, ideas and institutions within, across, below and beyond political borders. The approach can vary from micro, meso and to macro studies.

[46] The efforts included a temporary and specially designed exhibition of Norwegian industries at the Stock Exchange in Christiania [Oslo] and a special report on the Chinese visit and Norwegian businesses in China published in English by the leading Norwegian newspaper *Farmand*. Ship owners and other prominent merchants also hosted a lavish banquette in honour of the Chinese mission, where no expense seems to have been spared (Report of the visit of the Imperial Chinese Special Mission to Christiania, *Farmand* June 2, 1906).

However, at the core of all transnational studies is connectivity whether intentional or not, formal or informal, personal or impersonal, persistent or broken.

In this project, networks are used as analytical categories for such connectivity. However, stating the importance of human agency and networks is simple, analysing them as historical phenomena is not. There are two main reasons for this. Firstly, historical networks are constructs of historians trying to map human interactions and mutual dependencies of the past. This implies establishing a multitude of network typologies of formal and informal networks that co-exists and that may overlap. There were formal professional networks amongst allied missionary organisations operating in China, but clearly also informal contacts between missionaries in competing organisations. Similarly, the presence of formal shipping cartels was felt strongly in the China trade; however, there was also collusion through clandestine networks as well as informal lobbying against the power of the cartels. Moreover, different types of networks were likely to overlap, especially business networks and social networks.

Secondly, scholars tend to disagree on the importance of networks in the sense of whether humans would act differently within a network or not. Stephen Vertovec argues that there is a tendency to overemphasise the social capital of networks in terms of loyalty and what has been labelled as enforceable trust (Vertovec 2009: 51). The maritime and economic historian Gordon Boyce on the other hand, emphasises that networks contribute towards creating two competitive sets of rationalities within business; a communicative rationality and a calculative rationality. Boyce's main argument is that there was a special bond, a social capital within the maritime industries that spurred its agents to act according to communicative rationality as opposed to opportunism and calculative rationality (Boyce 2001, 2003, 2008). As opposed to transaction-cost economics, Boyce stresses that such networks can provide transaction benefits as opposed to transaction costs, which in turn emulates if not exceeds the benefits of corporate hierarchies in manufacturing industries. Preliminary results of network studies of Norwegian ship owners operating in China suggest, however, that the ship owners were immensely adaptable in their behaviour.

A prime example of this can be found through the life and experiences of Haakon Johan Wallem (1870-1951). Wallem was born in Bergen and received his business training in Germany. After short stays in London and Vladivostok, he ventured into shipping in China at the turn of the 20th century. Decades later, he was a leading figure in shipping in Shanghai and Hong Kong. Looking into Wallem´s contacts during the first phase of his businesses in China, i.e. 1903-1913, he had extensive networks in his hometown, in Germany and London as well as in Shanghai and

Hong Kong. A closer study of Wallem's business activities from this period reveals that he was able to act according to both communicative and calculative rationalities. The decisive factor in this respect seems to be the basis upon which the networks were founded, their location and context. Preliminary findings suggest that Wallem depended on networks forged at his hometown of Bergen for his investments, whilst a successful access to the multinational web of shipping firms, agents and agencies in Asia was the key to his success in the freight market.[47] The crucial point is how Wallem adapted to both contexts. Reports and court records from the Norwegian consulate-general in Shanghai, tell a story of numerous commercial disputes involving Wallem and mostly with Chinese or Japanese counterparts.[48] However, this should not be interpreted as being equivalent to the way Wallem operated in China as the nature of the sources only deals with conflicts, and hence does not capture the many business transactions that went smoothly. However, the sources suggest that calculative rationality was an integral part of business – also in the marine world. Wallem's network in Norway was primarily a horizontal one and rested upon a co-operative ethos and social capital. Turning the argument on its head, one could of course argue that communicative rationality is in reality also an expression of calculative rationality if one wishes investments and business to be repetitive games. Hence, in this respect Vertovec's suggestion that one should not exaggerate the social capital of networks appears to be a valid reminder not to rush to conclusions or stretch too eagerly for generalisations.

However, the picture becomes even more complex when looking at data available through the *North China Herald* on merchant shipping in Shanghai. To a maritime historian, the North China Herald holds a goldmine of information. The newspaper published shipping statistics on a weekly basis from the mid-19th century and up to 1916. Here, we find information on dates of arrivals and departures, names and nationality of vessels, names of captains, specifications of cargo, trade routes, time spent at sea – and we find names of the consignees for each of these journeys. As part of the merchants and missionaries project, the author of this chapter is currently constructing a database covering all vessels that were recorded having travelled through the port of Shanghai for a selected number of years for all nationalities.

[47] In the archives after the ship owner Jacob Christensen in Bergen, there is a string of letters and telegrams between Haakon Wallem and Jacob Christensen, mainly from 1910 and 1911. Characteristically, their content tends to be a blend between business and private correspondence. Bergen Maritime Museum Archival Collection, MS 0016, Jacob Christensen, Box D. *Correspondence.*

[48] For the purpose of this chapter, the reports and court cases involving Wallem prior to 1913 have been consulted. These can be found at the National Archives of Norway, Ministry of Foreign Affairs, S-2611, Utenriksstasjonene, Generalkonsulatet i Shanghai, Kina, UD 1905, Boxes 235 and 367. The cases include the dispute between Wallem and the Chinese citizen and compradore Mo Zu Mai (1907-1908), Fong Tai and Co. (1907-1911) a Japanese company registered in Shanghai and the Chinese citizen and former close business associate, Tong Yues (1910-1916). The dates here are referring to the duration of the disputes.

One of the ambitions of this project is to map maritime networks beyond the transnational sphere of Norway and China and to study how war acted as an agent of change within the business.

Figure 2.2 illustrates a clear tendency for consignees to give preference to vessels flying the same flag as themselves. This may partly be explained by the fact that many of these consignees acted as both brokers, agents and ship owners, and hence were chartering their own vessels. However, looking at the magnitude of the national preference, this clearly does not represent the full explanation. This hypothesis is also confirmed by looking at the market structure and concentration in the consignment market. The top three firms were the same for all the years observed here; Butterfield & Swire (UK), China Merchant´s Steam Navigation Company (China), and Jardine Matheson & Co. (UK). These three companies had a joint market share varying between 50 to 60% for all years observed here.

However, Figure 2.2 suggests a weaker correlation between the nationality of the consignees and the nationality of the vessels in 1904 and in particularly in 1905, than for 1894 and 1914. The most likely explanation behind this is the mobilisation of Japanese vessels for war purposes. A detailed look at the data reveals that Japanese consignees such as Mitsui Bishi Co, Nippon Yusen Kaisha and Mitsui Bussan Kaisha chartered foreign and predominantly Norwegian ships during the years of war. The Russo-Japanese War became a defining moment for several Norwegian merchants – including Haakon Wallem. The earnings from this war not only provided him with sufficient capital to venture from brokering into ship-owning as well, it also laid a foundation for future business through personal and business networks in China as well as Japan. Although this is obviously an area where further research is much

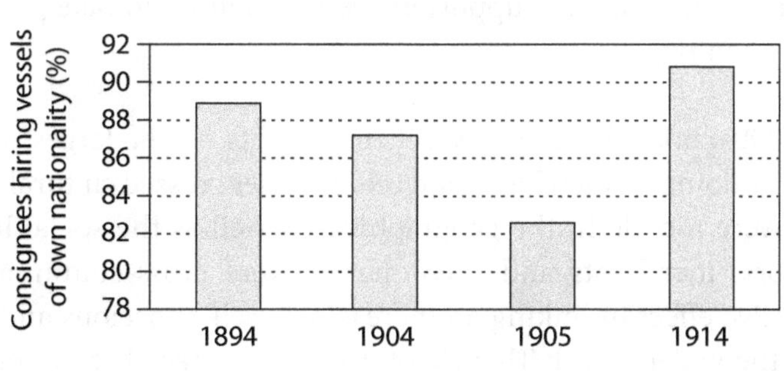

Figure 2.2. Vessels and consignees in the port of Shanghai 1894, 1904, 1905 and 1914 (Various issues of North China Herald. Number of observations: 1894 (2,527 voyages), 1904 (3,802 voyages), 1905 (4,165 voyages) and 1914 (4,699 voyages). Number of consignees: 1894 (52), 1904 (64), 1905 (74), 1914 (61)).

needed, these preliminary findings tend to suggest that networks at home played a significant role within the merchant world of shipping in China during the period covered here.

Concluding remarks

The relationship between China and the West in the late 19th and early 20th centuries has been at the epicentre of several overlapping – and interacting – academic discourses over the past decades. In terms of world history, traditional Eurocentric narratives are becoming fewer and the demand for a new history has been increasing. Some of the most vibrant expressions of this new history have come through global history. Global history has become a *genre* of comparisons and connectedness in time and space, whilst globalisation studies have a thematic focus on the dynamics of macro integration. At the same time, China is also in need of a new history of its own – perhaps especially with regards to its place within and relationship with the wider world. Odd Arne Westad is amongst those who have been the most vocal in the call for a broader approach in order to capture processes that shaped most people's experiences with foreign and international relations. It is our ambition to make the ongoing project *Merchants and Missionaries. Norwegian encounters with China in a transnational perspective, 1890-1937* a small, but helpful contribution in this direction.

At the heart of the project lies a desire to uncover and communicate largely forgotten experiences of when merchants and missionaries were nodes of connections between China, Norway and the wider world. Norway was a small country and a geopolitical minnow, but in the late 19th and early 20th century it played a disproportionate role in shipping and was an integral part of the world-wide-web of protestant mission. Around 1880, Norway had the third largest merchant fleet in the world and missionaries and religious supporters were mobilised to take part in the great missionary awakening.

In the early 1890s, both these groups set their sights on the largest market in the world: China. By doing so, merchants and missionaries crossed cultural and linguistic borders, and were mobile in the geographical as well as the social landscape. As historical actors, merchants and missionaries were cross-community migrants with a cumulative effect in linking an increasingly self-conscious and nationalistic Norway with the wider world. This chapter has suggested that migration theories may offer a missing link in existing studies of the mobility and impact of merchants as well as missionaries – especially through the perspectives of cross-community migration. Still, vertical and horizontal social mobility was only possible if allowed by prevailing institutions in terms of norms, attitudes and organisational setups. Addressing merchants and missionaries together also implies that women are integrated into a broader history of transnational encounters and history of globalisation. Recent research suggests that migrating women had a relatively higher potential of traversing pre-existing limitations to their social status. Networks offer a key to further our understanding of both geographical and social mobility. The strong transnational character of both business and mission implies that it is necessary to move within and beyond borders in order to grasp a fuller picture of these movements. For instance, whilst five missionary organisations in Norway took up work in China between 1889 and 1905, a similar number of US-Norwegian organisations were set up within the very same timeframe. Individual missionaries moved along and across these organisational boundaries at home, in transit as well as at the destination. However, whilst Vertovec reminds us that the social capital of networks should not be over-emphasised, Boyce has suggested that networks in the marine world create a special ethos in terms of trust and loyalty. Preliminary results of studies of merchant networks in Shanghai indicate that they may both be right. Networks have to be understood within a wider context, and so must the people within and outside such networks. Research into merchant networks with Shanghai as their hub is currently ongoing. The findings so far suggest that merchants were able to act according to communicative as well as calculative rationalities, and that they were immensely adaptable in their behaviour. Hence, networks appear to define a room of manoeuvre, but not necessarily a pre-destined pattern of behaviour.

It is hoped that this project will translate into analyses with relevance to the genre of global history and the topic of globalisation, both often characterised by top-down processes and meta-level generalisations. Such approaches too easily ignore the complexity, historicity and importance of human agency. Hence, by putting human agency and interaction at the very heart of our research, we will hopefully gain a better understanding of the transnational connectivity between China, Norway and beyond.

Acknowledgements

The author would like to thank the Norwegian Research Council for funding of the work presented here through Project Number 205553, Christian Guth for excellent research assistance and Malin Gregersen, Olga Medvedeva, Karina Hestad Skeie and Lars Fredrik Øksendal for valuable comments.

Bibliography

Austin, A. 2007. *China's millions: the China Inland Mission and late Qing Society, 1832-1905.* Wm. B. Eerdmans Publishing Co., Grand Rapids, MA, USA/Cambridge, UK.

Bayly, C.A. 2004. *The Birth of the Modern World 1780-1914 – Global Connections and Comparisons.* Blackwell Publishing, Oxford, UK.

Becker, B. 2010. Coastal Shipping in East Asia in the Late Nineteenth Century. *Journal of the Royal Asiatic Society Hong Kong Branch* 50: 245-302.

Bergen Maritime Museum Archival Collection, MS 0016, Jacob Christensen, Box D. *Correspondence*, Bergen, Norway.

Bickers, R. 1999. *Britain in China: Community, Culture and Colonialism, 1900-49.* Manchester University Press, Manchester, UK.

Bickers, R. 2004 [2003]. *Empire Made Me: An Englishman Adrift in Shanghai.* Allen Lane/ Penguin Books, London, UK.

Bickers, R. 2011. *The Scramble for China: Foreign Devils in the Qing Empire, 1832-1914.* Allen Lane/Penguin Books, London, UK.

Boyce, G.H. 2001. *Co-operative structures in global business: communicating, transferring knowledge and learning across the corporate frontier*, Routledge, London, UK.

Boyce, G.H. 2003. Network knowledge and Network routines: Negotiating Activities between Ship owners and Shipbuilder. *Business History* 45: 52-76.

Boyce, G.H. 2008. Network Structures, Processes and Dynamics: Inter-firm Cooperative Frameworks in the Shipping Industry. In: Fischer, L.R. & Lange, E. (eds.). *International merchant shipping in the nineteenth and twentieth centuries. The comparative dimension.* Research in maritime history no. 37, International Martime Economic History Association, St. Johns, Canada: 165-178.

Brandt, L., Ma D. & Rawski, T.G. 2013. From Divergence to Convergence: Re-evaluating the History Behind China's Economic Boom. *Economic History Working Papers* no. 175/13. Centre for Competitive Advantage in the Global Economy, Department of Economics, University of Warwick, Warwick, UK.

Brautaset, C. 2002. *Norsk Eksport 1830-1865 i perspektiv av historiske nasjonalregnskaper.* PhD thesis, Norwegian School of Economics, Bergen, Norway.

Brautaset, C. 2011. When Distance Matters: The output of the Norwegian Merchant Fleet, 1830-1865. In: Fischer, L.R. & Lange, E. (eds.). *New Directions in Norwegian Maritime History.* International Maritime Economic History Association, St. Johns, Canada: 37-56.

Brautaset, C. & Grafe, R. 2006. The Quite Transport Revolution: Returns to Scale, Scope and Network Density in Norway's Nineteenth-Century Sailing Feet, *Discussion Papers in Economic and Social History*, no.62, University of Oxford, UK.

Brautaset, C. & Tenold, S. 2008. Waiving the Rules to Rule the Waves – Globalisation and Norwegian Shipping Policy, 1850-2000. *Business History,* 2008, 50(5): 565-582.

Brautaset, C. & Tenold, S. 2010. Lost in Calculation? Norwegian Merchant Shipping in Asia, 1870-1914. In Fusaro, M. & Polónia, A. *Maritime History as Global History*. Research in martime history no. 43, International Martime Economic History Association, St. Johns: 203-222.

Brenner, R. 1976. Agrarian Class Structure and Economic Development in Pre-industrial Europe. *Past and Present* 70: 30-74.

Brunero, D. 2009 [2006]. *Britain's Imperial Cornerstone in China. The Chinese martime customs service, 1854-1949*. Routledge, Oxon, London, UK.

Cassel, P. 2010. Traktaten som aldrig var och föredraget som nästan inte blev. De svensk-norsk-kinesiska förbindelserna 1847-1909. *Historisk Tidskrift* 130(2): 437-466.

Chang, C. 2013. *Government, Imperialism and Nationalism in China: The Maritime Customs Service and its Chinese Staff*. Routledge, Oxon, London, UK.

Clossey, L. 2006. Merchants, migrants, missionaries, and globalization in the early-modern Pacific. *Journal of Global History* 1: 41-58.

Crossley, P. 2008. *What is Global History?* Polity Press, Cambridge, UK.

Dannevig, B. 1982. *Skipene fra Norge. Fortellingen om den gamle norske skipsfart på Det Fjerne Østen – særlig Kina og Øyene i Ostindia*. Nordanger Forlag. Bergen, Norway.

Dirlik, A. 2000. Is there History after Eurocentrism? Globalism, Postcolonialism and the Disavowal of History. In: Dirlik, A., Bahl V. & Gran, P. (eds.). *History after the Three Worlds. Post-Eurocentric Historiographies*. Lamham, Boulder, Rownman & Littlefields Publishers, New York, USA/Oxford, UK: 25-47.

Ekeland, T., Anderson. A. & Christensen O.T. (eds.) 1919. *White unto harvest: a survey of Lutheran United Mission, the China mission of the Norwegian Lutheran Church of America*. The Board of Foreign Mission, Augsburg Publishing House, Minneapolis, MN, USA.

Fairbanks, J.K. & Goldman, M. 2006 [1992]. *China – A New History*. 2nd enlarged Ed., Kelknapp Press of Harvard University Press, Cambridge, MA, USA/London, UK.

Farmand, Report of the visit of the Imperial Chinese Special Mission to Christiania. June 2 1906.

Fenby, J. 2008. *The Penguin History of Modern China. The fall and Rise of a Great Power, 1850-2008*. Allen Lane, London, UK.

Frank, A.G. 1998. *ReOrient: global economy in the Asian Age*. University of California Press, Berkeley, CA, USA.

Gills, B.K. & Thompson, W.R. 2006. *Globalization and Global History*. Routledge, London, UK/ New York, USA.

Goldstone, J.A. 1993. *Revolution and Rebellion in the Early Modern World*. University of California Press, Los Angeles, USA/London, UK.

Goldstone, J.A. 2008. *Why Europe? The Rise of the West in World History 1500-1850*. McGraw Hill, New York, NY, USA.

Grafe, R. 2010. Turning maritime history into global history. Some conclusions from the impact of globalisation in early modern Spain. In: Fusaro, M. & Polónia, A. *Maritime History as Global History*. Research in martime history N 43, International Martime Economic History Association, St. Johns, Canada: 249-266.

Hannerz, U. 1996. *Transnational Connections: Culture, People, Places*. Routledge, London, UK/New York, NY, USA.

Helle, I. 2012. *Med verdifull last. Norske rederiers deltagelse i frakt av kinesiske arbeidsmigranter i perioden 1895-1930*. MSc thesis, University of Bergen, Bergen, Norway.

Hoerder, D. 2002. *Cultures in Contact: World Migration in the Second Millennium*. Duke University Press, Durham, UK.

Hsiao, L-L. 1974. China's Foreign Trade Statistics, 1864-1949. *Harvard East Asian monographs* 56, Harvard University Press, Cambridge, MA, USA.

Jones, E. 2004 [1981]. *The European Miracle: Environments, Economics and Geopolitics in the History of Europe and Asia*. 3rd Ed. Cambridge University Press, Cambridge, UK.

Jones, G. 2000. *Merchants to Multinationals. British Trading Companies in the nineteenth and twentieth centuries*. Oxford University Press, Oxford, UK.

Kirby, W.C. 1995. China, Unincorporated: Company Law and Business Enterprise in Twentieth Century China. *Journal of Asian Studies* 54(1): 43-63.

Klovland, J.T. 2002. Business cycles, commodity prices and shipping freight rates: Some evidence from the pre-WWI period. *SNF Report* no. 48/02. Institute for research in economics and business administration, Bergen, Norway.

Kocka, J. 2003. Comparison and Beyond. *History and Theory* 42(1): 39-44.

Landes, D. 1999 [1998]. *The Wealth and Poverty of Nations*. Abacus, London, UK.

Lucassen, J. & Lucassen, L. 2009. The mobility transition revisited, 1500-1900: what the case of Europe can offer to global history. *Journal of Global History* 4(3): 347-377.

Manning, P. 2005. *Migration in World History*. Routledge, London, UK.

Manning, P. 2006. Cross-Community Migration: A Distinctive Human Pattern. *Social Evolution & History* 5(2): 24-54.

Marks, R. 1998 [1997]. *Tigers, rice, silk, and silt: environment and economy in late imperial south China*, Cambridge University Press, Cambridge, UK.

Marks, R. 2002. *The origins of the modern world: a global and ecological narrative*. Rowman & Littlefield, Lanham, MD, USA.

Martell, L. 2007. The Third Wave in Globalization Theory. *International Studies Review*, 9(2): 173-196.

Martell, L. 2010. *The Sociology of Globalization*. Polity Press, Cambridge, UK.

Mikaelsson, L. 2003. Med frigjørende kunnskap til kvinner i Kina: misjonæren Hanna Holthe og hennes skole i Yiang. In: Hellesund, T. & Okkenhaug, I.M. (eds.). *Erobring ogoverskridelse. De nye kvinnene inntar verden 1870-1940*. Unipub Forlag, Oslo, Norway: 161-178.

Minuti, R. 2012. Oriental Despotism. *European History Online* (EGO), published by the Leibniz Institute of European History (IEG), Mainz, Germany, 2012-05-03. Available at: http://www.ieg-ego.eu/minutir-2012-en.

Myklebust, O.G. 1980. *H.P.S. Schreuder. Kirke og Misjon.* Gyldendal Norsk Forlag AS, Oslo, Norway.

Myrstad, I. 2009. *Generalkonsulatet i Kina. En studie av en svensk-norsk unteriksstasjon 1842-1905.* MSc thesis, University of Bergen, Bergen, Norway.

National Archives of Norway, Ministry of Foreign Affairs, S-2611, Utenriksstasjonene, Generalkonsulatet i Shanghai, Kina, China. UD 1905, Box 235.

National Archives of Norway, Ministry of Foreign Affairs, S-2611, Utenriksstasjonene, Generalkonsulatet i Shanghai, Kina, China. UD 1905, Box 367.

Osterhammel, J. & Petersson, N. P. 2005, *Globalization. A short history.* Princeton University Press, Princeton, NY, USA.

Pels, P. 2009. Missionaries. In: Iryie, A. & Saunier, P.-Y. (eds.). *The Palgrave Dictionary of Transnational History.* Palgrave Macmillan, Basingstoke, Hampshire, UK: 716-719.

Pomeranz, K. 2000. *The great divergence: China, Europe, and the making of the modern world economy.* Princeton University Press, Princeton, NY, USA.

Pomeranz, K. 2001. Is There an East Asian Development Path? Long-Term Comparisons, Constraints, and Continuities. *Journal of the Economic and Social History of the Orient,* 44(3): 322-362.

Reed, J. 1983. *The Missionary Mind and American East Asia Policy 1911-1915.* Harvard, Harvard East Asian Monographs 104: Harvard University Press, Cambridge, MA, USA.

Remer, C.F. 1933. *Foreign Investments in China.* The MacMillan Company, New York, NY, USA.

Sannes, K.O., Grandhagen, E., Hegertun, T., Jørgensen, K, Norseth, K. & Olsen, R. (eds.) 2008. *Med Kristus til Jordens Ender; festskrift til Tormod Engelsviken.* Tapir Forlag, Trondheim, Norway.

Saunier, P.-Y. 2009. Transnational. In Iryie, A. & Saunier, P.-Y. (eds.). *The Palgrave Dictionary of Transnational History.* Palgrave Macmillan, Basingstoke, Hampshire, UK: 1047-1055.

Scholte, J.A. 2005 [2000]. Globalization: A Critical Introduction. 2nd Ed. Palgrave Macmillan, Basingstoke, Hampshire, UK.

Schreiner, J. 1963. *Norsk skipsfart under krig og høykonjuktur 1914-1920.* Norges Rederiforbund, J.W. Cappelens Forlag, Oslo, Norway.

Seeberg, S. & Filseth, G. 2000. *I Yangzidragens rike: nordmenn i Shanghai gjennom 150 år.* Schibsted, Oslo, Norway.

Sejersted, F. 1993. *Demokratisk kapitalisme.* Universitetsforlaget, Oslo, Norway.

Spence, J.D. 1999 [1990]. *The Search for Modern China.* 2nd Ed. W.W. Norton & Company.

Strandenæs, T. (ed.). 2006. *Misjon og Kultur: Festskrift til Jan-Martin Berentsen.* Misjonshøgskolens forlag, Stavanger, Norway.

The North China Herald, various issues 1894, 1904, 1905, 1914. Available at: http://www.brill.nl/publications/online-resources/north-china-herald-online.

Thelen, D. 1999. The Nation and Beyond: Transnational Perspectives on United States History. *The Journal of American History* 86(3): 965-975.

Tyrrell, I. 2010. *Reforming the World*. Princeton University Press, Princeton, NY, USA.

Van Kessel, I. (ed.) 2002. *Merchants, Missionaries and Migrants. 300 Years of Dutch-Ghanian Relations*. KIT Publishers, Amsterdam, the Netherlands.

Vertovec, S. 2009. *Transnationalism*. Routledge, Oxon, London, UK.

Von Mende, E. 1971. *Die wirtschaftlichen und konsulären Beziehungen Norwegens zu Chinavon der Mitte des 19. Jahrhunderts bis zum 1. Weltkrieg*, Inaugural-dissertation zur Erlangung des Doktorgrades Der Philosophischen Fakultät der Universität zu Köln. Universität zu Köln, Cologne, Germany.

Wendt, H. 2011. *Die Missionarische Gesellschaft. Mikrostrukturen einer kolonialien Globalisierung*. Franz Steiner Verlag, Stuttgart, Germany.

Westad, O.A. 2012. *Restless Empire: China and the world since 1750*. Bodley Head, London, UK.

Wilkins, M. & Schrøter, H.G. 1998. *The Free-standing company in the world economy 1830-1996*. Oxford University Press, Oxford, UK.

Wong, B. 2000. *China Transformed: Historical Change and the Limits of European Experience*. Cornell University Press, New York, NY, USA.

Ytreberg, N. 1951. Fart og farvann. 1880-1914. In: Worm-Müller, J. (ed.). *Den Norske Sjøfarts Historie. Fra de Ældste Tider til Vore Dage*. II Bind, Tredje del. J.W. Cappelens Forlag, Oslo, Norway: 271-286.

Chapter 3: Opening the gates

China's Maritime Customs Service (1854-1949) in the contemporary context

Olga Medvedeva

In recent years, China has become a key centre of global trade. According to the World Trade Organisation (WTO), China is now the world's largest exporter and second-largest importer (WTO trade statistics, 2013). The process of the rapid integration in the world economy started in 1978, when Deng Xiaoping claimed the policy of reform and 'opening-up'. The growing role of the 'awakening dragon' is widely discussed among economists, political scientists and military experts (Huang & Zhou 2012, Nie *et al.* 2012, Ross 1992, Wortzel 2013). Nevertheless, there is a tendency to forget the historical experience before 1949, when China opened the 'gates' towards the world market, the outward and inward flows of goods, services, capital, people and knowledge. In the 19th century, the country experienced the general decline of the last Chinese Qing Dynasty (1644-1912). The state was suffering natural calamities, economic problems, and popular uprisings, the bloodiest of which, the Taiping Rebellion, took lives of over 20 million people (Heath & Perry 1994). At the same time, the country became an arena of conflicts between European, American, Japanese, Russian and Chinese interests. A rage of so-called 'unequal treaties' signed with foreign powers, stimulated the creation of the treaty port system. The first one was the treaty of Nanjing of 1842. It was the main result of the First Opium War (1839-1842) between British and Qing empires. The major ports became open for international trade, Western residents got the right of extraterritoriality and established their concessions, and China lost the customs tariffs autonomy.

Established in 1854, the foreign inspectorate of the maritime customs was a part of this system. This new administration represented a good example of the institutional interaction between this Asian country and the rest of the world in the late 19th/ early 20th centuries. This period is often associated with the first wave of global economic integration (Baldwin & Martin 1999). The external involvement has lost its significance in the mainland China after 1949, when the Communist Party came to power. From 1854 to 1949, the maritime customs had cosmopolitan character (over 20 nationalities were involved in the service but predominantly Chinese and British), and was focused on the foreign seaborne (and later overland as well) trade revenue

C. Brautaset and C.M. Dent (eds.) **THE GREAT DIVERSITY**
DOI 10.3920/978-90-8686-786-8_3, © Wageningen Academic Publishers 2014

collection.[49] New and reorganized customs stations were established among the coast and expanded to interior. The service played an important role in global trade development. This institution is of a special interest when studying China from the perspective of transnational history. In general terms, customs service can be defined as a government agency that is responsible for border management and collects duties from the international trade (the main part of which belongs to the seaborne commerce). Customs tariffs directly affect the relative prices on the international market, situation in the domestic economy, as well as the competiveness of countries and companies. The character of this institution might differ from one country to another, changes over time and have its own historicity.

Many works devoted to the history of the Chinese maritime customs during the period from 1854 to 1949 are written by both the Chinese and Westerners (Aitchison 1983, Cai 1989, Hansson 1946, King 1930, Wang 1969). Some of them wanted to share their experiences through the genre of memoirs, and others aimed at providing academic accounts on various topics. Historical writings published after 1949 in mainland China are mostly based on the Marxist methodology that stresses aggressive imperialism and semi-colonial China. The most prominent writer is Cheng (1993, 1999). Two main frameworks can pointed within the Western literature: the concept of the informal empire and an idea of the synarchy. The first one describes the Chinese maritime customs as a 'Britain's imperial cornerstone'. This approach is based on the British dominance in the service management and the 'pseudo-colonial presence' (leased territories, numerous settlements, substantial for China investments), and in a recent academia is supported by historians Robert Bickers (1999) and Donna Brunero (2006). The second one was proposed by American sinologist John King Fairbank, who considers the creation of the foreign inspectorate of the maritime customs as some sort of Sino-Western compromise. According to him, this phenomenon is based on the previous historical experience. Since the 4th century, alien rule (meaning the dynasties of barbarian conquest), such as Wei (386-556), Khitan Liao (907-1125), Jurchen Jin (1115-1234), Mongolian Yuan (1206-1368) and Manchurian Qing was a recognized practice in the Chinese political life (Fairbank 1953). The latter works, in its turn, tend to emphasize the malignancy of the Chinese bureaucracy and progressive role of the foreign-dominated institution.

This chapter aims to bridge the gap between Western and Chinese literature by studying Chinese contemporaries' views on the maritime customs, and a special focus is made on the period of the foreign involvement. The purpose is to trace how it

[49] There was also Native Customs Service [*changguan* 常关] that was collecting revenue from domestic trade，and from 1901 to 1931 was managed by the foreign inspectorate of the maritime customs (Tsai Weipin 2008).

was perceived and explained by the people of the same time, how the interpretations of the service mirror the context of its historians, and how these ideas influenced on the institutional development. It is not an easy task to study almost 100 years history of the foreign-managed customs service, and moreover how challenging it is to track the thoughts regarding this government agency in the works of the Chinese scholars, and further to discover the deep layers of their worldview.

In order to better comprehend the essence of this institution and the role of its traditional elements, the analysis begins with the main characteristics of the maritime customs before the establishment of the foreign inspectorate in 1854, and followed the development of the new administrative system until its separation into general administration of customs in the People's Republic of China and the customs of the Republic of China on Taiwan in 1949. The study is based on the selective works that were written by the Chinese scholars before 1949.

Chinese Maritime Customs before 1854: trade development and sea ban 'under Heaven'

The Chinese modern word for the 'customs' *haiguan* 海关 is composed of two characters. The first *hai* 海 means 'sea', the second *guan* 关 has quite illustrative origin. According to the Chinese dictionary *Shuowen Jiezi*, the hieroglyph *guan* means 'to keep the door closed with a wooden stick' [*yi mu heng chi menhu ye* 以木横持門戶也].[50] Thus, the etymology of the Chinese term 'customs' can be understood as 'to keep the sea frontiers closed'. It tells about the priority of the defensive responsibilities. For comparison, in the English language the origin of that word is more connected with the fiscal functions. According to the etymological dictionary written by Eric Partridge, the 'customs' is 'merely the plural of *custom* in its obsolescent sense', 'a duty expected by ruler or government' (Partridge 1958: 136).

Chinese historical writings claim that the fortified military guarding posts named *guan* were established on the overland trade routes as early as during the Western Zhou Dynasty 西周 (1046-771 BC) (Cai 1989). These border stations can be traced in the Chinese classic literature. Mencius 孟子 (372-289 BC), one of the most influential ideologists of Confucianism, in the work *Jin Xin II* 盡心下 refers to the checkpoints *guan* in the ancient society that was usually understood in the Chinese tradition as a role model:

[50] *Shuowen Jiezi* 说文解字 [The Explanation of Simple and the Analyse of Compound Characters] is one of the first Chinese dictionaries from the Han Dynasty 汉朝 (206 BCE-220 CE).

> Anciently, the establishment of the frontier-gates was to guard against violence. Nowadays, it is to exercise violence (Mencius, translated by James Legge 1895: 481).[51]

In another work *Gong Sun Chou I* 公孫丑上, the same author describes the ideal policy towards the customs tariffs:

> If, at his frontier-passes, there be an inspection of persons, but no taxes charged on goods or other articles, then all the travellers of the kingdom will be pleased, and wish to make their tours on his roads. … Then the people in the neighbouring kingdoms will look up to him as a parent. … Thus, such a ruler will not have an enemy in the entire kingdom, and he who has no enemy in the kingdom is the minister of Heaven (Mencius, translated by James Legge 1895: 200).[52]

Thus, the border checkpoints were used as an instrument to control the flow of the people and goods, and included the combination of the two main tasks: to protect from the external threat and to promote the internal prosperity. In practical terms, they also functioned as tax collectors, although official ideology of Confucianism promoted the free trade principles. The first 'sea checkpoints' that were called *haiguan* 海关, appeared in the 17th century. One of the main known specialised source on the history of that institution before the First Opium War is the *Annals of the Guangdong Maritime Customs* [*Yue Haiguan Zhi* 粤海关志]. It was finished by the scholar Liang Tingnan 梁廷枬 (1796-1861), and was published in 1839.

In order to better understand the way how the customs service was perceived in the traditional society, it is important to mention a special ideological framework that is reflected in this tractate. In the Chinese traditional cosmology, the Son of Heaven [meaning Emperor] is holder of *mingling* 命令 [the mandate of Heaven] and rules the *Tianxia* 天下 [Celestial Empire, often translated as 'all under Heaven']. The term *Tianxia* refers to the entire geographical world and the territory under emperor's control. Thus, China was considered as the centre of the civilization, and the tributary states 'clustered on its borders' (Li 2002: 26). Customs checkpoints were a part of this worldview. They were supposed to support the right order in the *Tianxia*, and regulate the relations between Emperor and remote vassals (Liang 1839).

The point of departure of the analysis in the *Annals of the Guangdong Maritime Customs* is the role of the designated places for the overland trade *hushi* 互市, and

[51] The Scottish sinologists James Legge (1815-1897) translated the Chinese term *guan* as the 'frontier-gates'.

[52] In this passage, the Chinese word *guan* is translated by J. Legge as the 'frontier-passes'.

so-called 'ship-trade' *shibo* 市舶 that already existed during the Han Dynasty (206 BC-220 AD). The same source mentions such institution as the Maritime Trade Bureau [*shibo shi* 市舶使, later renamed *shibo si* 市舶司] that was founded during the Tang Dynasty 唐朝 (618-907). This centralized service represented an integrated part of the traditional administrative system of the Six Ministries[53], was responsible for the seaborne trade control, and can be considered as a prototype of the customs service *haiguan*. Significant changes came during the Ming Dynasty明朝 (1368-1644), when the court started to apply so-called 'sea ban' policy, which was practiced until the early Qing Dynasty. Under this regime ordinary people were not allowed to participate in import-export activities that were limited to the tribute trade.[54] As a result, the Maritime Trade Bureau lost its practical importance.

In 1684, the Kangxi Emperor 康熙帝 (1654-1722) issued an order 'to lift the sea ban' (Liang 1839). Thereby, the four customs stations *haiguan* were founded in Shanghai [*Jiang* 江], Ningbo [*Zhe* 浙], Zhangzhou [*Min* 闽], and Guangzhou [*Yue* 粤]. [55] The treaty of Nanjing implies that exactly the above-mentioned areas should be open to the British merchants. The naval customs, the same as the Maritime Trade Bureau earlier, was controlled by the court. Superintendents were usually appointed from the Imperial Household Department [*nei wu fu* 内务府] and the Ministry of Revenues, very often with the help of the local governors or military leaders (Cai 1989). It was closely connected to the Ministry of Rites, the Ministry of War, the Ministry of Justice, and Office of Barbarian Control [*li fan yuan* 理藩院]. The coastal defence officers [*haifang guan* 海防官] played an important role in that service (Liang 1839).

Modern scholars have different opinions concerning the court's international trade policy, as well as various ideas about Chinese isolationism, and the attitude towards the sea. For example, Charles Frederick Remer described the Chinese economy as self-sufficient: it did not depend on the import until opium was introduced (Remer 1967). Li Kangying analysed the phenomenon of the Chinese isolation during the Ming Dynasty. He pointed not only such aspects as the threat from the piracy and the fear of military conflicts with other counties, but also internal pressure and

[53] The system of the Six Ministries was the central administrative system in China since ancient time, and consisted of a Ministry of Personnel [*li bu* 吏部], a Ministry of Revenue [*hu bu* 户部], a Ministry of Rites [*li bu* 礼部], a Ministry of War [*bin bu* 兵部], a Ministry of Justice [*kang bu* 刑部], and a Ministry of Works [*gong bu* 工部] (Hucker 1958).

[54] Tribute trade is often described in the recent literature as a part of the China-centric world order, where vassal states were supposed to provide the curtain tribute to the Chinese court, and could exchange this 'gift' for the Chinese goods (Rowe 2009).

[55] The place named Zhangzhou 漳州 borders Xiamen 厦门. According to the Chinese scholar Cheng Shiqi, the responsibility for the *Min* Customs 闽 was distributed between the ports of Xiamen and Fuzhou. By 1766, the clearance was conducted in Xiamen, but Fuzhou military general was involved in the station control (Cheng Shiqi 1993).

the influence of the Confucian ideology, particularly its 'anti-commercial moral foundation' and cultural supremacy (Li 2010: 25). John King Fairbank considered China before the 19th century as a continental empire that 'accustomed to foreign relations across a land frontier' (Fairbank 1953: 33). Historian Zeng Yangwen is more focused on the importance of the maritime factor (Zheng 2012).

Some of the reasons, why China tended 'to keep the door closed' in the early 19th century, are mentioned in the contemporaries' works. In the *Open Letter Addressed to the Sovereign of England* can be traced fear of opium danger and lack of interest in foreign products. This work was written in 1839 by the Chinese scholar and civil servant Lin Zexu 林则徐 (1785-1850) who is well-known for the fight against the drug trade. He threatened to close the domestic market for imports in case the British continued to bring opium, and also pointed out that Chinese goods are very valuable for foreigners, while the latter take into China 'nothing more than baubles' (Lin Zexu, quoted in Senin and Yang 1961: 55-56).

The foreign inspectorate in the late Qing Empire (1854-1912)

The establishment of the foreign inspectorate of the maritime customs service can be considered as a direct response to unrest in the country and external pressure from the Western countries. In 1853, during the Taiping Rebellion, the Small Sword Society [*Xiao Dao Hui* 小刀会] revolted in Shanghai, seized the customs house, and made it difficult for the local authorities to collect international trade revenue. It was a golden opportunity for foreign merchants to influence the institutional framework of their activities in China. It also gave a possibility for treaty powers to secure the funds to pay indemnities that China was obliged by the agreements. In 1854, the Shanghai official *daotai* 道台 Wu Jianzhang 吴健彰 (1791-1866) and three consuls – British (Sir Rutheford Alcock), American (Robert Creighton Murphy) and French (Benoit Edan) – agreed that each of these three countries delegated one representative, called the 'the tax affairs superintendent' *shuiwu jiandu* 税务监督 [in English version 'inspector of customs']. They should control the revenues collections from overseas trade. New management showed its effectiveness. In 1859, the foreign inspectorates became an integral part of the Chinese administrative system and got the name of the 'tax officer' *shuiwu si* 税务司 [in English 'customs commissioner']. They gained a central place in the service. It is quite remarkable that the Chinese name of the new inspectorate was linked to the origin of the British customs service in terms of the derived name – collection of duties. The whole institution was called the same Chinese term as before 1854 – *haiguan*. Until the overthrow of the Qing dynasty in 1912, the English name of this new administrative system was the Imperial Maritime Customs Service (IMCS).

After 1861, when *Zongli Yamen* 总理各国事务衙门, the office that was in charge of the foreign affairs was established, the new customs administration was placed under its control. It was an unprecedented shift from the integral element of the traditional administrative system of the Six Ministries to the newly arranged department responsible for the relations with other countries. Thus, the control over the international commerce gained more 'external' character. How was it possible that the Qing court delegate the rights to collect the trade revenue to the Westerners that traditionally were associated with 'barbarians'[56]? How can this 'alien takeover' be explained by the Chinese contemporary context?

China's defeat in the Opium Wars, first of all, revealed the West's technical and military superiority. During the 1840s-50s, the scholar Wei Yuan 魏源 (1794-1856) advocated for improvement and development of the maritime defence in his multi-volume work, the *Illustrated Treatise on the Maritime Kingdoms* [*Haiguo Tongzhi* 海国图志]. He also propagandized the slogan 'Learn the superior technology of the barbarian, in order to control him' [*shi yi zhi zhang ji zhi yi* 师夷之长技以制夷]. In the 1861, Prince Gong 恭亲王 (1833-1898) who was regent during the reign of the young Emperors Tongzhi 同治帝 (1856-1875) and Guangxu 光绪帝 (1871-1908), and the influential statesman Li Hongzhang 李鸿章 (1823-1901) initiated the Self-Strengthening Movement [*ziqiang yundong* 自强运动]. This range of reforms is also known as the Westernisation Movement [*yangwu yundong* 洋务运动], and lasted until 1895. The original goal was to adopt the best Western scientific achievements in order to strengthen the empire. Traditionally, only the Confucian *Four Books* [*Si Shu* 四书] formed the basis for officially recognised education. In order to implement foreign knowledge into the existing ideas, Chinese thinkers developed their own methodological basis through the principle *zhongxue wei ti, xixue wei yong* 中学为体，西学为用. In direct translation: Chinese disciplines are the basic, and the Western science serves the practical purposes. This idea provided possibility to combine traditional morality and Western efficiency. Modernisation reforms started in the military sector, followed by the railway, mining, telegraph and marine development sectors, expanded to educational system, and was accompanied by translation of Western literature of different fields.

The IMCS became a part of this movement. The new customs administration did not only show an effective management (the service rapidly covered treaty ports along the coast, expanded to inland, and ensured the growing revenues collection), it was also significantly engaged in the technical upgrading. Its activities expanded

[56] According to the traditional thoughts, Emperor got power from the Heaven to rule the world [*Tianxia*], and China is a centre of 'human universe'. Those who were not a part of Chinese culture, we considered 'barbarians' (Li 2002).

to harbour and waterways management, meteorological observations, postal service development, just to mention some of them. Moreover, a part of the customs revenues was used for reforms development. The reformer Li Hongzhang, who respected Western achievements generally, also highly valued the IMCS contribution to the development navigation technology in China:

> China's maritime domain is extensive. Ports and rivers branches are scattered and stretched out for more than ten thousand *li*. Inspector General Hart spent years in order to establish the patrolling ships service, lighthouses, floating piles and more than 260 other things. … The Inspector General Hart put his heart and soul in this work, and should be awarder as encouragement (Li Hongzhang, quoted in Cheng 1993: 298).[57]

Thus, the initiators of the reforms appear not to have perceived foreign involvement in the customs service as the threat to the Qing Empire. Paradoxically enough, the Chinese government took advantage of Western management by using it to help solve domestic problems, to support the development of overseas trade, as well as to keep the balance in relations with other counties. The longevity of foreigners in the IMCS may have a multitude of explanations. One of these explanations rests upon the assumptions that the foreigners possessed skills and expertise that their Chinese counterparts could not match. It was the reforms that allowed the service to implement 'external' elements in the government system. The treaty system stimulated the process of the opening of the Celestial Empire. The 'sea ban' policy and strategy of 'keeping the door closed', which were practised during the Ming and the early Qing periods, proved ineffective against the Western technical superiority by the 19th century. The inevitability of the clash with the 'barbarians' caused the issue of the proper tactics. The process of modernisation with foreign participation can be considered as one of the best solutions: to harness their power in order to become stronger.

However, many Chinese intellectuals, who supported the reforms and were relatively familiar with the situation in the Western countries, were not pleased with the existing rules for the overseas trade. They were particularly concerned with customs tariffs. In 1843, British and Qing Empires signed a treaty that fixed the import-export duties. In 1858, the treaty of Tianjin established the principle of a 5% *ad valorem* tax on both imports and exports. This imposed unilateral change in the customs taxation system represented a loss in China's tariff autonomy. It was only regained much later, between 1928 and 1930. How was it possible that Chinese authorities agreed on such

[57] Author's translation from Chinese.

obligations? In 1930, the Chinese scholar and civil servant Wang Jingchun[58] 王景春 (1882-1956) published *How China Recovered Tariff Autonomy*. He referred to the traditional Sino-centric ideology, and argued that the court in Beijing, as well as all politicians, were only familiar with the classical Confucian dogmas. Hence, their education did not offer clear ideas about the rest of the world, and they could not properly understand the meaning of 'tariff autonomy' (Wang 1930).

The customs tariffs policy became of a special attention after 1876, when imports began to prevail over exports (Remer 1967). Ma Jianzhong 马建忠 (1844-1900) studied the international law in the Paris Institute of Political Studies from 1876 until 1880. Upon his return to China, he became involved in the modernisation movement. According to him, foreign trade was an essential source of wealth, and customs autonomy was very crucial for China (Ma, translated by Bailey 1998). Another well-known scholar who stressed the importance of the customs regulations was Zheng Guanying 郑观应 (1842-1922). He came from a comprador[59] background and mastered Western languages. In his famous work *Words of Warning in Era of Prosperity* [*Shengshi Xin Weiyan* 盛世新危言], published in the 1890s, he looked at the customs tariffs as the main weapon in the economic competition, and applies to the experience of other countries:

> According to the overseas regulations, the import of the goods is considered as an obstruction to the national well-being, and is a subject of the heavy duties. The export of the native products is lightly taxed: it stimulate the sale and easies the people's hardship later edition of (Zheng 1988: 574). [60]

He encourages the ombudsmen to pay more attention on the recent situation in the overseas trade that he describes as the 'commercial war' between China and foreign powers:

> Disaster of the military conquest is easily perceived. It is like wind that blows out the lamp. The death of the state coursed by the commercial takeover is invisible. It is like oil that gives out, and let the lamp expire Zheng 1988: 561).[61]

Both Zheng Guanying and Ma Jianzhong supported a trade policy built upon principles of protectionism. They argued that the government should protect the Chinese domestic economy and to enhance the competitiveness of Chinese products on the world market. Ma Jianzhong suggested dividing the import goods into several

[58] His name is often written as Wang Ching-Chun.

[59] Historian Hao Yen-p'ing describes comprador as 'Chinese manager of a foreign firm in China, serving as middleman in the company's dealings with Chinese' (Hao Yen-p'ing 1970: 1).

[60] Author's translation from Chinese.

[61] Author's translation from Chinese.

sub-groups, which are supposed to form the basis for taxation. According to him, the authorities should differentiate the import tax according to the product group's importance (varying from 5 to 30% *ad valorem*), and reduce the export tax. It is worth noting, that after 1930, the new tariffs system implied very similar taxation principle: 7.5-27.5% *ad valorem* on imports (depending on the type of goods), the export duties were in average around 4% (Lebedinskaya 1958).

The main advocates of free trade policy were Chen Chi 陈炽 (1855-1900), who was an active participant in the Hundred Days' Reform [*Wu Xu Bian Fa* 戊戌变法] of 1898, and Yan Fu[62] 严复 (1854-1921). Chen Chi hoped to use Western economic ideas to revive the ancient Chinese ideal society. According to him, Britain's embrace of Adam Smith's ideas on the virtues of free trade had lead Britain to be 'crowned the whole world' (later edition of Chen 1997: 149). Yan Fu believed in the concept of the social and trade liberalization. The Western principle of freedom is claimed vital for the country's prosperity and development. It stimulates the personal responsibility for the state affairs, as well as competition between people and counties (Yan Fu, in Senin and Yang 1961).

In practical terms, the IMCS was not a place of prestige amongst the locals. First, it can be explained by the still existing suspiciousness towards non-Chinese things. In 1894, the reformist and scholar Tan Sitong 谭嗣同 (1865-1898) wrote *A Reply to Bei Yuanzheng* [*Bao Bei Yuanzheng* 报贝元征]. He criticised the neglect against everything that has foreign character, including the work in the customs service:

> During the last decades Chinese scholarly noblemen have been talking only about the principles of morality, and worried only about their status. They think that learning 'overseas matters' dishonours them. Therefore, they oppose it in every way. People that had some sense of shame did not dare to have any relations with foreigners. That is why only miserable people that do not worry about their reputation, want to be involved in the 'overseas matters'. Some of them started as servants and achieved the rank of customs officers (Tan Sitong, quoted in Senin and Yang 1961: 235). [63]

Second, the carrier opportunities were very limited for the Chinese in the IMCS. Another active supporter of the reforms Yung Wing[64] 容闳 (1828-1912) was the first Chinese who was sent to the United States to study. After his graduation from Yale College in 1854, he worked in the translation department of the IMCS. In

[62] Yan Fu is well-known as the translator of the Thomas Huxley's *Evolution and Ethics*, Adam Smith's *Wealth of Nations* and Herbert Spencer's *Study of Sociology*.

[63] Author's translation from the Russian version of the Chinese original text.

[64] The normative spelling of his name is Rong Hong, Yung Wing is the version in the Cantonese dialect.

1909, he published the book *My Life in China and America*, where he emphasized the corruption, very limited carrier possibilities and the prepossession towards 'Chinaman' in the service:

> ...being an educated man, I saw no reason why I should not be given the same chances to rise in the service of the Chinese government as an Englishman, nor why my individuality should not be recognized and respected in every walk of life... I did not care to associate myself with a pack of Custom-house interpreters and inspectors, who were known to take bribes (Yung 1909: 64-65).

Implementation of the foreign elements in the maritime customs occurred against the background of the general decline of the Qing dynasty. The supporters of the Self-Strengthening Movement realized the difficult situation in the country. The original aim of the reforms before the republic period was to use Western knowledge to maintain the traditional social and political regimes. The learning of the 'overseas matters' reshaped the traditional picture of the world, and opened the way to new disciplines. There was also initiated the translation of the Western works on economics, which included information about modern financial and tax systems. In 1886, IMCS supported the publication of *The Theory of Political Economy* written by the British economist William Stanley Jevons. The Chinese version was named *Fu Guo Yang Min Ce* 富国养民策 [*The Suggestions for How to Support People and Make the Country Rich*]. During 1902-1907, the international trade policy and the modern Western economic theories were actively discussed in the newspaper called *New Citizen* [*Xin Min Cong Bao* 新民叢報].[65] This journal was published in Japan; the main editor was well-known scholar and reformer Liang Qichao 梁启超 (1873-1929).

After administrative reforms of 1906, the position of the IMCS in the government system gradually started to change. It was not under *Zongli Yamen* [the office of foreign affairs] control anymore, but moved under jurisdiction of the newly established Tax Office [*shui wu chu* 税务处]. The IMCS fulfilled its tasks in the modernisation process, and started its deep integration into the domestic administrative system. In 1908, in Beijing was founded the first Taxation School [*shuiwu xuetang* 税务学堂, later renamed *shuiwu zuan xuexiao* 税务专门学校] that provided modern education for the Chinese customs officers. The first graduates and potential successors of foreign employees appeared in 1912 (Cai 1989).

[65] As example can be mentioned a series of articles: *Waiguo Maoyi Lun* 外国贸易论 [*The Discussion about the Foreign Trade*] and *Jin Shi Yingguo Shangye Zhengce zhi Fazhan* 近世英国商业政策之发展 [*The Development of the Modern British Commercial Policy*] by Zhong Yuan 重远 (see in the later edition of Liang 2008).

The Chinese Maritime Customs Service and nationalistic sentiments in republican China (1912-1949)

The Xinhai revolution [辛亥革命] of 1911, which overthrew the last Qing Dynasty and led to the establishment of the Republic of China in 1912, had a significant impact on the historical role of the customs service. The official English name of the institution was changed into the Chinese Maritime Customs Service (CMCS). Its development was accompanied by the dramatic political changes and the volatility in the country: the presidency of Yuan Shikai 袁世凯 (1912-1915), warlord period (1916-1927), the so-called Nanjing decade (1927-1937), the Sino-Japanese War and the confrontation between the Guomindang and the Communist Party (1937-1949). The decentralization of power and the widespread turmoil after the revolution caused difficulties in many areas – also in the tax collection. However, at the same time a weak central control gave opportunities for the foreign inspectorate to gain more influence. The CMCS became an important institution that saved its integrity and could guarantee revenues, as well as secure and manage China's overseas loans. Financial control made the service a significant tool in the Western hands until the 1927, when it became the subject of the Office of Customs Affairs [*guanwu shu* 关务署] of the Ministry of Finance [*caizheng bu* 财政部] in the National Government (Sun 2005).

Wang Jingchun famously argued that 'the revolution of 1911 taught China that to set up a republic cost more money than to continue the old monarchy' (Wang 1930: 270). The young republic needed money. Hence, the status and importance of the service elevated accordingly. Increasing familiarity with Western theories and deepening national identity made the problem of the existing overseas trade policy more acute: why did other countries have the privilege of tariffs autonomy, but not China? The focus on CMCS issues transformed into a voluminous production of works during the 1910s and 1920s. This was a new generation of scholars. Most of them were economists that had received education abroad, mainly in the United States, and after return to China worked in the civil service, or became influential politicians. They looked at the customs service through the frame of modern political and economic theories, as well as studied the different institutional systems in Western countries. The main purpose of these writings can be explained by pragmatic interests, namely to gain tariffs autonomy and to get rid of what was perceived as unwanted foreign interference.

One of the first works to offer such a scientific explanation was written by Chen Weiping 陈维屏 (1897-?). In 1915, he graduated from Boston University. The thesis was titled *The Development of the Chinese Customs during Qing Dynasty: 1644-*

1911, and provides a detailed analysis of the institution's evolution. Zhu Jin 朱进 (1886-?) graduated with a doctorate from Columbia University and argued that the existing customs revenue system was a disaster for the national economy and should be revised (Zhu 1919). Advocating free trade, he believed that the economic development should be based on growing consumer demands. Sun Yatsen 孙中山 (1866-1925), 'the father' of the Republic of China, stressed the importance of economic factors for political sovereignty. The cancellation of the 'unequal treaties' were considered by him as crucial to national development. In 1924, in the interview to *The Japan Chronicle* he said:

> ...there are the two main things that should be done: the abolition of the foreign extraterritoriality and the recovery of the customs autonomy (later edition of Sun 2006: 424-425).[66]

He argued that the Chinese government had the same rights to be independent as the leaders of European countries, and hence should be able to make important decisions concerning the customs regime. Sun Yatsen compared economic policies of various countries, and came to the conclusion that the state support was necessary: particularly, to increase import fees, to abolish the export tax and inner taxation *lijin* [often called '*likin*' 厘金] (Sun 2006).

In 1925, Yang Desen 杨德森 published his work entitled *The Evolution of the Chinese Customs System* [*Zhongguo Haiguan Zhidu Yange* 中国海关制度沿革*I*]. This study, as well as Cheng Weiping's aforementioned work, refers to a long history of the institutional independence. The loss of the latter is explained by the weak Qing government that signed the range of the 'unequal treaties'. The CMCS is described as a bureaucratic structure, based on the British administrative system and conquered by Westerners. In 1926, Chen Xiangyuan 陈向元 claimed that the customs service after 1854 was developed under military pressure from foreign powers. He appealed for a revision of the tariffs system, and urged China to pay the necessary attention to her financial and economic situation:

> After I started my deep study of the recent economic and political issues and their origins, I was totally shocked how dangerous they are. One of the scariest problems is the bondage of the customs tariffs. We must reform this system. The doom of the Chinese finance is not allowed (Chen 1926: preface).[67]

Jia Shiyi 贾士毅 (1887-1965) completed his studies at Meiji University in Japan in 1911, and later became a secretary of the Taxation Department in the Ministry of

[66] Author's translation from Chinese.

[67] Author's translation from Chinese.

Finance. In 1929, he published a book *Customs Tariffs and the National Sovereignty* [*Guanshui yu Guoquan* 关税与国权], where he studied key economic issues that became a vital concern for China during the Nanjing decade. These academic works contained argumentations underpinned by strong nationalistic sentiments. The authors were influenced by their own national context and were moreover influential. After 1930, when the Nanjing government gained the full customs tariffs autonomy, it adopted a protectionist trade policy.

In the 1920s, it is possible to trace a new tendency: a view on the service as a tool against the imperialistic aggression, especially after Guangzhou government failed in the gaining control over customs revenues during 1923-1924. In 1924, in the interview given to the *Guangzhou Daily*, Sun Yatsen explained the crucial role of the proletariat in the solution of national problems. He remarked that 'imperialistic powers insist on the low tariffs and control the CMCS; it makes the industrial working class weak' (Sun 2006: 40). A similar attitude can be found in the work *China under Economic Aggression* [*Jingji Qinlüe Xia zhi Zhongguo* 经济侵略下之中国], written in 1925 by Qi Shufen[68] 漆树芬 (1892-1927). In 1927, the first edition of the CMCS magazine for Chinese staff *Guan Sheng*[69] 关声 was issued in Shanghai. The magazine raised a wide range of topics, from smuggling to global customs policies and life in the service. The *Guan Sheng* signalled an increased focus on the many inequalities suffered by Chinese staff, and demanded better treatment for the native personnel. After 1928 when National government united control over the country, the growing patriotic sentiments resulted in the changes in the structure of employment: Chinese officers started to appear within middle and upper level ranks of the service. The customs service was not perceived as an important part of the westernisation reforms or the disgraceful place to work anymore, as it was typical during the late Qing dynasty, but as a platform for the struggle of national interests.

Conclusion

Historical writings confirm that China's maritime customs has a long history. The seaborne trade control was organized as early as the 7th century. Institutional development had meanwhile a rather undulating character, with periods of rise and decline. Its evolution was closely connected to the attitude towards overseas trade and the threat from the outside world. The establishment of the foreign inspectorate of the maritime customs in 1854 was understood in the Chinese works written before 1949 as the continuity of the service rather than a break. Such a view was

[68] Qi Shufen 漆树芬 was a student of the famous Japanese Marxist economist Hajime Kawakami 河上肇 (1879-1946) from Kyoto Imperial University.

[69] *Guan Sheng* 关声 can be translated as 'the voice of the customs service'.

not unreasonable. First, the Nanjing treaty of 1842 implied the opening of the same ports, where traditional customs stations were functioning since the 17th century. China 'opened the gates' to the international trade in the literal sense of this phrase, as the Chinese word for the 'customs' means 'to keep the sea frontiers closed'. Second, the institution kept the same Chinese name *haiguan* (which is still in use, and is applied to both naval and overland customs dealing with the foreign trade and cross-border movement of goods). Third, officially the new customs administration was under Chinese government jurisdiction.

The peaceful conversion of the maritime customs from the part of the traditional system of the Six Ministries to the Western-organised civil service and integrated part of the government system happened thanks to the modernisation reforms. This was initiated by the Qing court in 1861, and aimed to use Western (primary technical) achievements in order to strengthen the empire. Chinese scholars worked out the crucial methodological basis in order to implement new knowledge: Chinese studies are the foundation, and Western knowledge serves the practical purpose of improvement and refinement. Foreign involvement in administration became an integral part of this process, had limited ideological influence, and perfectly fitted into the tactics of this approach. It's primary goal was more effective tax collection functionality. The chosen strategy was more like a 'soft fist' style in the martial arts: its philosophy of victory implies the principle of maximum use of the enemy's force. Influential scholar Wei Yuan called to learn from 'barbarians' in order to control them not so long before the foreign inspectorate was established. Initially, 'foreign help' was skilfully applied by the local authorities to solve current problems. After national leaders gained enough power, Western participation in the CMCS was viewed as unnecessary.

After reaching a net import position in 1876, progressive scholars started to worry about the existing customs system that imposed 5% *ad valorem* tax on both imports and exports. Although traditional Confucian doctrine advocated free trade, there were intellectuals that perceived the existing situation as a 'trade war' and advocated protectionist customs duties. The first wave of the educated abroad Chinese intellectuals together with Westerners brought to the Celestial Empire not only technical knowledge, but also modern Western economic, social and politic theories that were transforming the world system. The concept of *Tianxia* (Celestial Empire, often translated as 'all under Heaven') started to co-exist with the idea of nation. The Chinese realized that the relations between countries might be based on the principle of sovereignty but not on tributary relations, and that customs tariff autonomy was an international norm. After the establishment of Republic of China in 1912, the CMCS became an important source of tax revenue, and tariffs autonomy turned out to be a vital issue for the Nationalist government. It was a then new generation of intellectuals that were decided the future of China's overseas commerce. They contributed not only to the service's development, but also to a wider ideological transformation within the society. The recovery of the customs tariffs autonomy after 1930 allowed the Chinese authorities to pursue a protectionist trade policy.

Thus, the Chinese contemporary context shows that the perception of the maritime customs changed significantly over time. In traditional society it was recognized as the institution that supports the right order in *Tianxia* by regulating commercial relations between the Emperor and remote vassals. After 1854 it was perceived as a controversial yet effective agent for modernisation, and later became a vehicle for anti-imperialistic struggle. This chapter stresses the multifaceted role of the foreign customs inspectorate in Chinese history and how this varied from one period to another. Furthermore, it has been shown that from an intellectual historical perspective there was no universal approach to studying the transformation of China's customs service. At the same time, we can see the way how the historical experience of the previous wave of global economic integration is applied in this Asian country in recent years. Since 1978, foreign trade has been played a crucial role in the Chinese economic development. China would not be able to become a key centre of global trade without a comprehensive and robust customs service. Today, this government agency represents an institution with centralised control, uses modern technology and skilled native staff, and ultimately protects key development-related national interests.

Acknowledgements

This study was funded by the Norwegian Research Council, the Project Number 205553.

Bibliography

Aitchison, J. 1983. *The Chinese Maritime Customs Service in the Transition from the Ch'ing to the Nationalist Era: An Examination of the Relationship between a Western-style Fiscal Institution and the Chinese Government in the Period before the Manchurian Incident*, PhD thesis, University of London, London, UK.

Baldwin, R.E. & Martin, P. 1999. Two Waves of Globalisation: Superficial Similarities, Fundamental Differences. *NBER Working Paper no. 6904, January 1999*. National Bureau of Economic Research, Cambridge, MA, USA.

Bickers, R.A. 1999. *Britain in China: Community Culture and Colonialism, 1900-1949*. Manchester University Press, Manchester, UK.

Brunero, D. 2006. *Britain's Imperial Cornerstone in China: The Chinese Maritime Customs Service, 1854-1949*. Routledge, New York, NY, USA.

Cai, W. 1989. 中国海关简史 [*A Brief History of the Chinese Customs Service*]. Zhongguo Zhanwang Chubanshe, Beijing, China.

Chen, C. 1997. 续富国策: 自叙 [*Implementation of Policies to Enrich Counties: Autobiography*]. Zhonghua Shuju, Beijing, China.

Chen, X. 1926. 中国关税史 [*The History of the Chinese Customs Tariffs*]. Jinghua Yinshuju, Beijing, China.

Cheng, S. 1993. 中国近代海关史:晚清部分 [*The History of the Modern Chinese Customs Service: The Late Qing Period*]. Renmin Chubanshe, Beijing, China.

Cheng, S. 1999. 中国近代海关史:民国部分 [*The History of the Modern Chinese Customs: The Republic Period*]. Renmin Chubanshe, Beijing, China.

Fairbank, J.K. 1953. *Trade and Diplomacy on the China Coast: the Opening of the Treaty Ports, 1842-1854*. Harvard University Press, Cambridge, MA, USA.

Hansson, P.C. 1946. *25 År i Midtens Rike* [*25 Years in the Middle Kingdom*]. Erindringer, Oslo, Norway.

Hao, Y.-p. 1970. *The Comprador in Nineteenth Century China: Bridge between East and West*. Harvard University Press, Cambridge, MA, USA.

Heath, I. & Perry, M. 1994. *The Taiping Rebellion 1851-66*. Osprey Pub Co, London, UK.

Huang, W.C. & Zhou, H. 2012. *Dragon versus Eagle: the Chinese Economy and U.S.-China Relations*. W.E. Upjohn Institute for Employment Research, Michigan, MI, USA.

Hucker, C.O. 1958. Governmental Organization of the Ming Dynasty. *Harvard Journals of Asiatic Studies* 21: 31-36.

Jia, S. 1929. 关税与国权 [*Customs Tariffs and the National Sovereignty*]. Shangwu Yinshuguan, Shanghai, China.

King, P.H. 1930. *In the Chinese Customs Service: A Personal Record of Forty-seven Years*. H. Cranton, London, UK.

Lebedinskaya, L.N. (ed.). 1958. *Istoriya Economicheskogo Razvitiya Kitaya 1840-1948: Sbornik Statisticheskikh Materialov* [*The Chinese Economic History 1840-1948: The Selected Statistic Materials*]. Izdatelstvo Inostrannoi Literatury, Moscow, Russia. Russian translation of the Chinese original work *Zhongguo Jindai Jingji Shi Tongji Ziliao Xuanji* 中国近代经济史统计资料选辑. 1955. Kexue Chubanshe, Beijing, China.

Legge, J. 1895. *The Chinese Classics: With a Translation, Critical and Exegetical Notes, Prolegomena and Copious Indexes, Works of Mencius*. Clarendon Press 2, Oxford, UK.

Li, K. 2010. *The Ming Maritime Trade Policy in Transition, 1368 to 1567*. Harrassowitz Verlag, Wiesbaden, Germany.

Li, Z. 2002. Traditional Chinese World Order. *Chinese Journal of International Law* 1(1): 20-58.

Liang, Q. 2008. 新民叢報 [*New Citizen*]. Zhonghua Shuju, Beijing, China.

Liang, T. 1839. 粤海关志 [*Annals of the Guangdong Maritime Customs*]. Available at: http://ctext.org/library.pl?if=gb&res=2390&remap=gb.

Ma, J., translated by Bailey P.J. 1998. *Strengthen the Country and Enrich the People: the Reform Writings of Ma Jianzhong*. Curzon Press, Richmond, Surrey, UK.

Nie W., Dowell, W. & Lu, A. 2012. *The Shadow of the Dragon: The Global Expansion of Chinese Companies – How It Will Change Business Forever*. AMACOM, New York, NY, USA.

Partridge, E. 1958. *Origins: A Short Etymological Dictionary of Modern English*. Routledge & Kegan Paul, London, UK.

Qi, S. 1925. 经济侵略下之中国 [*China under Economic Aggression*]. Guanghua Shuju, Shanghai, China.

Remer, C.F. 1967. *The Foreign Trade of China*. Ch'eng-Wen Publishing Company, Taipei, China.

Ross, H.M. 1992. Awakening Dragon. The Real Danger in Asia is from China. *Policy Review* 62: 10-17.

Rowe, W.T. 2009. *China's Last Empire: The Great Qing*. Harvard University Press, Cambridge, MA, USA.

Senin, N.G. & Yang, X. (eds.) 1961. *Izbrannye Proizvedeniya Progressivnikh Kitaiskikh Myslitelei Novogo Vremeni (1840-1898)* [*Chinese Progressive Thinkers of Modern Times: Selected Works (1840-1898)*]. Akademiya Nauk SSSR, Moscow, Russia.

*Shuowen Jiezi*说文解字 [*The Explanation of Simple and the Analyse of Compound Characters*]. Available at: http://www.shuowen.org/index.php?pinyin=guan&p=3.

Sun, X. 2005.中国近代海关史大事记 [*The Main Events in the History of the Chinese Modern Customs Service*]. Zhongguo Haiguan Chubanshe, Beijing, China.

Sun, Y. 2006. 孙中山全集 [*Complete Works of Sun Yatsen*]. Zhonghua Shuju 11, Beijing, China.

Wang, C.C. 1930. How China Recovered Tariff Autonomy. *Annals of the American Academy of Political and Social Science* 152: 266-277.

Wang, W. 1969. 滥竽海关四十年 [*Forty Years' Experience in the Customs*].

World Trade Organisation. Statistics Reports. Available at: http://www.wto.org/english/res_e/statis_e/its_e.htm.

Wortzel, L.M. 2013. *The Dragon Extends It's Reach: Chinese Military Power Goes Global*. Potomac books Inc, Dulles, VA, USA.

Yang, D. 1925. 中国海关制度沿革 [*The Evolution of the Chinese Customs System*]. Shangwu Yinshuguan, Shanghai, China.

Yung, W. 1909 *My Life in China and America*. Henry Holt Company, New York, NY, USA.

Zheng, G. 1988. 郑观应集 [*The Collected Works of Zheng Guanying*]. Renmin Chubanshe 2, Shanghai, China.

Zheng, Y. 2012. *China on the Sea. How the Maritime World Shaped Modern China*. Brill, Leiden, the Netherlands.

Zhu, J. 1919. 中国关税问题 [*The Chinese Customs Tariffs Issues*]. Wen Hai Chubanshe, Taibei, China.

Chapter 4: The world wide missionary web

Transnational mission networks' influence on the 19th and 20th century Norwegian China missions

Karina Hestad Skeie

> Transnational networks among ... missionaries generated systems of information gathering and dissemination a century before the invention of the World Wide Web.
>
> *(Tyrell 2010: 40-41)*

In the years just before and after 1890, and simultaneously in Sweden, Norway, and Scandinavian immigrant milieus in the United States, individuals who called themselves 'friends of the Chinese', came together and founded associations and mission societies specifically aimed at Christian mission in China. Urged by what they experienced as a divine and personal calling to become Christian missionaries to China, numerous men and women sought to the newly founded societies. In 1890 alone, five different mission organisations from Norway, Sweden, Denmark, and Scandinavians in North-America, sent men and women to work as Christian missionaries in central China and Manchuria (Claesson 2001, Kjebekk 1991, Tiltnes & Vig 1946). More were to follow.

The Nordic China-fervour did not take place in a vacuum. It was part and parcel of the wider Protestant mission movements' interest in- and interaction with China, which started rather modestly in 1807 with Robert Morrison (1782-1834) of the London Missionary Society (LMS). Towards the turn of the century, particularly from the 1880s onwards, Protestant missions in China grew steadily in size, strength and reach (Reinders 2004). On the one hand, the Protestant missions' expansion in China was closely connected to British-Chinese merchant-relations (Bays 2012) and USA's increased interest in China from the turn of the 20th century (Tyrell 2010). On the other hand, it was part of a world-wide Protestant mission movement, already working in many African countries, in the Middle-East, in India, Mongolia and other countries in Asia. United in a common and rather loosely defined vision of expanding 'God's Kingdom unto the ends of the world', Protestant Christians across national and denominational borders mobilized in formal and informal transnational networks of individuals, institutions, ideas, money and methods (Nielssen, Okkenhaug & Skeie 2011). Towards the turn of the 20th century, Protestant missions' international and trans-denominational cooperation and contact became

C. Brautaset and C.M. Dent (eds.) **THE GREAT DIVERSITY**
DOI 10.3920/978-90-8686-786-8_4, © Wageningen Academic Publishers 2014

more organised and more frequent. This was reflected, among other things, in a renowned World Missionary Conference in Edinburgh in 1910 (Stanley 2009) and the growth of transnational and trans-denominational organisations like the Student Volunteer Movement for Foreign Missions. It was also crucial in the emergence of the Norwegian China missions.

Transnationalism is a key concept for research in human and social sciences that focus on links and interactions of people and institutions across nation-state borders (Nielssen, Okkenhaug & Skeie 2011). Connected to processes of globalisation, the term 'transnational' denotes links and flows of people and ideas operating over, across, through, beyond, above and in-between polities and societies (Iriye & Saunier 2009). When I use the term 'transnational connections' rather than 'globalization', this is because 'transnational' is considered more suitable for phenomena of variable scale and distribution (Hannertz 1996). The transnational character of Christian Protestant missions is multifaceted and includes individuals, groups, movements and enterprises. Importantly, the term 'transnational' also allows for a parallel focus on the continued importance of the 'national' even as it is transcended. As Christian missions are connected to- and articulate with local and national processes both at home and abroad, the particular dynamic of organisations like missions stem in no little part from the fact that they are simultaneously national and transnational.

Understanding the transnational Protestant mission movement as a web of connections and networks, I draw on a definition of transnational collective action as '...the process through which individuals, non-state groups, and/or organizations mobilize jointly around issues, goals, and targets that link the domestic and international arenas' (Von Bülow 2010: 5). Von Bülow sees social networks both as a precondition of- and an outcome of collective action. This is because actors' pre-existing networks influence choice and movement, yet creates new links and connections which in turn restrain or further future action (Von Bülow 2010: 7). In order to explain cooperation and connecting bonds between actors, it is necessary to reveal more than their mere common interest. More important is how actors are able (or not able) to overcome differences and how they construct common purpose. This, Von Bülow argues, constitutes an '...agency-centred view of networks as the product of choices of their members and as processes of meaning attribution' (Von Bülow 2010: 7).

This chapter traces the beginning of the Norwegian China missions and how their missionaries and the mission societies were influenced by and connected to other Christian missions and religious milieus in Norway, Europe, the United States and China. I focus on the first Norwegian individuals associated with Christian mission in China and the three first Norwegian China-mission societies albeit concentrating mostly on the mission society called the Norwegian Lutheran China Mission Association [*Det norske lutherske Kinamisjonsforbund,* or NLCM].

Probing into formal and informal contact and interactions between the Norwegian China-missions and other missions and missionaries, I am interested in connection and mobility through time and space. Mobility depends on cultural and social capital. But it also depends on available technologies of movement. It is no coincidence that missionaries started moving to China in greater numbers as travel was democratized (Cresswell 2006) and improved technologies transformed trade, tourism, transport and communications into global rather than national, inter-European, and transatlantic pathways and networks (Tyrell 2010). How mobility is experienced, and how people move are closely connected to how mobility is represented and made meaningful (Cresswell 2006). Thus, one key to understanding individuals' and groups' motivation for the Christian mission endeavour in China (and elsewhere) is the importance of particular representations of the world, specific understandings of the mission and the processes through which these images and perceptions were experienced as expressions of divine intent compelling people to act. In order for the interrelated actions and activities to be set in motion locally and trans-locally, these sets of ideas and representations had to resonate with individuals and groups' notions and motivations.

China's millions': a powerful transnational imagery

The history of the Norwegian China-missions is both directly and indirectly connected to the prominent missionary James Hudson Taylor (1832-1905) of China Inland Mission (CIM), and his powerful representation of 'China's Millions' in dire need for the Christian gospel. The mission interest for China grew rapidly in Norway from the 1880s, allegedly both because CIM's mission-magazine *China's Millions* was read in Norway, and because Norwegian mission magazines translated and reprinted accounts from CIM's and Hudson Taylor's work (Kjebekk 1991). This underlines the thoroughly transnational and to some extent also trans-denominational character of the literary side of the Protestant mission movement (Hofmeyr 2005, Mikaelsson 2003). Furthermore, it shows the intimate interconnectedness between a particular imagery of China and individual and collective engagement and responses.

The mission literature's description of mission encounters was characterized by a mixture of 'documentary from the field' and spiritual reflection. It was aimed at specific readers and tuned to reach particular goals. As Gullestad commented:

> Within the evangelizing modality, in particular, missionary stories aim at evoking engagement with the spiritual and material suffering of people in Africa (and elsewhere), piety to God, as well as confidence in the religious teachings of the mission and in its ability to help by acting on God's behalf (Gullestad 2007: 21).

Within this framework, support of the mission, either by becoming a missionary, donating money, carrying out voluntary work or otherwise promoting the noble cause, were transformed to actions on God's behalf. Taylor, described as 'one of the great storytellers of the Victorian age' (Austin 2007: xxiii) was among the virtuoso of this religious genre. He could 'take a simple poignant anecdote that showed how God was working in his own life, and amplify and refine it until it gained mythic proportions' (*ibid.*). Tayolor's illustration, twelve million Chinese souls lost for God every year due to the lack of missionaries to preach the gospel to China's four hundred million inhabitants, was at once overwhelming and accessible. The mobilizing potential of this particular imagery is underlined through the fact that Hudson Taylor's book *China's Spiritual Needs and Claims* (1865) allegedly secured the initial funding of the China Inland Mission (Bundy 2009). Even more telling is nevertheless the vast number of missionaries that Taylor's book and imagery inspired to go to China.

The multifaceted manner in which the mission-encounter imagery worked to evoke engagement, piety, faith in spiritual teaching and the notion that the mission could – and did – act on God's behalf through the missionary work, may indicate why mission-interest and mission-agitation often went hand-in-hand with revivals and revivalist spirituality. The mission and the revivalist imagery and aims were overlapping and therefore also mutually reinforcing. Around 1900, the NLCM and its emissaries were the main catalysts of spiritual revivals on the west-coast and on the southern coast in Norway (Seland & Aagedal 2008). Also individual travelling revival evangelists at this time seem to have tended to combine missionary propaganda from China as part of their message to repent and revive. Some were themselves missionaries or missionaries-to-be, using preaching-tours as fund-raising for their own journey and work in China. One example is the former teacher Ole Næstegård (1861-1937) who first went to China on his own in 1888 and worked with the Swedish missionary Erik Folke (1872-1939). In 1890, he returned to Norway, among other things to raise more funds (Brandtzæg 1902).

Other evangelists, particularly some of the more prominent, were thoroughly transnational and trans-denominational in outreach as well as in education. Of great importance to the Norwegian China missions, was the Swedish-American Fredrik

Franson (1852-1908). Born in Nora, Sweden, Franson emigrated to the U.S.A. at 17, with two of his brothers. During a revival in 1874, Franson was baptized in a so-called Holiness Swedish-Baptist church in Nebraska. Subsequently he got educated as an evangelist at the renowned L. D. Moody's church in Chicago, and became an international evangelist and traveller (Bundy 2009). Franson volunteered to recruit missionaries for China after hearing of Hudson Taylor's call. Indeed, this recruitment became his 'life work' in Europe and North America (Bundy 2009).

The first Norwegian missionaries to China allegedly received their missionary calling through Franson's preaching. Two single young women, the governess Sofie Reuter (1860-1892) and housemaid Anna Jakobsen (1860-1913) worked as servants in merchant Theodor Rasmussen's home in Kristiansand. The Rasmussen couple housed Franson whenever he was in Kristiansand during his preaching-campaigns in Norway. Revival meetings were also held in the Rasmussen house. Since there was no Norwegian mission working in China at the time, Jakobsen and Reuter went out for the CIM in 1885 (Kjebekk 1991). The CIM was a faith-mission. This implied that the responsibility of securing financial support for missionary work was placed upon the individual missionaries themselves rather than their sending-society. Jakobsen and Reuter received money from Norwegian China-interested individuals and support groups, in particular from Norwegian Free Church milieus in their hometown of Kristiansand. In some instances, close and personal ties developed between particular support-groups and individual missionaries.

James Hudson Taylor visited the Norwegian capital Kristiania (today Oslo) in 1889. This resulted in the founding of the first Norwegian mission society which specifically targeted China. The Norwegian China Mission (NCM; later renamed the Evangelical Orient-mission) based in the Norwegian Free Church, was founded in 1889 as a Norwegian branch of the CIM (Bundy 2009). Already the same year, the first Norwegian NCM missionaries were sent to China, concentrating their work in the Shensi and Shanxi Provinces in northern China. Differently from other mission societies in Norway at the time, but in correspondence with the CIM, the NCM did not create an organisation with a board and a general assembly that collected money, selected and sent missionaries to the field, and made all major decisions for the mission both 'at home' and 'in the field'. In contrast to board mission-organisations, which were usually associated with denominational and national churches, the CIM, lived and promoted a more entrepreneurial organisation.

The principle of 'living by faith', that is, as simply and inexpensively as possible, and to place trust in God providing for their livelihood, appears in part to have developed from Taylor's hard-earned experiences during his first period as a missionary for the China Evangelization Society from 1854-1860 (Bundy 2009). When his mission society failed to send money for his support for months on end, he was forced to live in Chinese buildings, and eat like Chinese working class citizens (*ibid.*). Learning that this was in fact possible, Taylor came to appreciate the independence this represented vis-à-vis the missionary work. It liberated him from the decisions of more or less well-informed mission-leaders in Britain, who knew little of what (according to Taylor) was necessary in China, yet made decisions for his missionary work that he had to follow. This experience was transformed into a mission strategy that made living close both to God and to the people among whom they worked a spiritual ideal.

Local spiritual and organisational differentiation

When the CIM was founded within free-church milieus as the first Norwegian faith-mission in 1889, two prominent board mission societies already existed in Norway. The Norwegian Mission Society (NMS) was founded in 1863, and had worked in Zulu-Natal since 1848 and Madagascar since 1866. Madagascar was NMS' main and most successful mission-field, judged by speed and the number of converts (Skeie 2012). The Norwegian Santal Mission was founded in 1867 and worked among the Santal people in India, in close cooperation with British missionaries. In the following I will concentrate on the NMS, because out of the two oldest missionary organisations – only the NMS started up work in China.

Albeit an independent, voluntary organisation, the NMS had strong ties with the Norwegian Lutheran state church. This was evident in different ways. Firstly, state church pastors had prominent positions in the NMS boards and often also in the local mission associations. Secondly, the NMS' (male) missionaries who received a five-year theological education at the mission school in Stavanger, were ordained and inaugurated as mission pastors by state-church bishops prior to going to the mission fields. Thirdly, the missionaries, the NMS and the general Norwegian public understood the NMS to represent the Norwegian Lutheran state church in the countries where the NMS worked despite the fact that, as an independent voluntary organisation, the NMS was not a state church mission in a strict formal sense.

Support committees for Christian China mission were founded in many large cities influenced by religious revivals that were prominent in many Nordic countries at the time (Seland & Aagedal 2008). Towards the end of the nineteenth century, several support groups were started in cities like Kristiansand, Mandal, Trondheim, Ålesund, Kristiania, Stavanger and Bergen. These were support groups consisting of people belonging to Lutheran state-church milieus. In Bergen, the support committee was founded in 1890 on the initiative of a number of young, single professional women, following a particularly inspirational mission meeting, where the Norwegian CIM missionary-to-be Sivert Gjerde preached a message visibly inspired by Hudson Taylor's afore mentioned book. The powerful image of twelve million Chinese souls lost for God every year due to the lack of missionaries to preach the gospel to China's four hundred million inhabitants, incited this group of 30-40 women to act.

Due to the NMS' and the state church's prominent position, the Bergen support-committee for the China-mission invited the powerful NMS secretary Lars Dahle to a meeting in Bergen, in 1891. At the meeting, the China mission support-committee tried to convince the NMS to take up missionary work in China. However, Dahle and the NMS declined. It was evident that their main attention was still directed at Madagascar and Zulu-Natal. Before the meeting with China mission supporters in Bergen, the issue of a China-mission had already been discussed within the NMS organization itself; after all, the powerful image of 'China's dying millions' also triggered NMS supporters' feelings of responsibility. The internal arguments against starting up in China were both the NMS economy – one could not afford a third field at this point, and the fact that there were already some thirty Protestant mission societies working in China (Nome 1943: 108).

At the Bergen meeting with the China-friends, Dahle used his prominent position in Norwegian Christianity and society, to challenge the Lutheran orthodoxy of the China-mission support milieus in Norway. The CIM's non-denominational attitude was by many seen as a threat to true (Lutheran) Christianity. Dahle also criticised faith-missions for being irresponsible; sending individuals with little education to an ancient writing culture like the Chinese, and leaving the missionaries to take care of their own support often under difficult circumstances (Kjebekk 1991). Dahle also did not support a Norwegian China mission, claiming that a mission in China was comparatively much more expensive than in Zulu-Natal and Madagascar. This economical argument did not go well with the China interested support groups. Did Dahle imply that money should decide which souls to save, they asked provokingly (Kjebekk 1991).

According to the NLCM's own history, the Bergen committee members' dedication for the China mission cause was merely fuelled by what they experienced as Dahle's and NMS' condescending attitude (Kjebekk 1991). Despite Dahle's efforts at advising against it, the support committees for China mission allegedly felt compelled to act on their own and create their own mission, so powerful was the divinely inspired urge for China's millions. Underscoring their denominational affiliation, and at least indirectly responding to part of Dahle's denominational critique, the new society was named The Norwegian *Lutheran* China Mission Society. It was further decided that it should be a society that financially supported the missionaries and the missionary work in China. The NLCM was thus *not* a faith mission, in contrast to the NCM and the CIM. Already in October 1891, initiative was taken to gather the many China-support groups and to create a new Norwegian mission society that was to send Norwegian missionaries to China. The NLCM was founded in 1891, and sent out its first eight missionaries to China the same year: two women and six men. The NLCM took up work in Hupei and Honan in Central China, in Shensi and eventually also in Manchuria, Kunming and Kweiyang in south-western China. Today the mission society, renamed The Norwegian Lutheran Mission Association [*Norsk Luthersk Misjonssamband,* or NLM] is among the largest mission societies in the Nordic countries.

The fact that the transnational upsurge to send Christian missionaries to China resulted in several new Norwegian mission societies reflected an increasing religious differentiation already in motion in the Norwegian society. A law issued in 1845, the so-called Dissenter Law, had for the first time made it legal to leave the Norwegian Lutheran State Church and to found faith communities and denominations outside the state church. The state church itself harboured a range of different views and practices within its religious landscape.

Since the NLCM and the NMS drew supporters largely from the same religious milieus in Norway, there was a clear element of competition between the two societies from the start. The religious difference between the NMS and the NLCM was also in the lay-dominated and low-church orientation of the NLCM's leaders, and in the fact that the great majority of its male missionaries were not Lutheran pastors (in contrast to the situation in the NMS). The difference between the NMS and the NLCM was further accentuated as the Norwegian state church refused NLCM's request to ordain the mission's only theological candidate, Johannes Brantzæg, before he was sent off to China (Kjebekk 1991). The NLCM from then on carried out its own sending ritual in prayer- and assembly houses around Norway (Kjebekk 1991).

That both men and women in the NLCM received the same missionary-inauguration, and that both sexes were called 'missionaries' from the start, was radical both in the mission and in Norway at the time. In comparison, the NMS reserved the missionary title for men well into the 20th century (Skeie 2012). The organisational differentiation between the NLCM and the NMS thus made visible existing differences regarding the missionary role, missionary education and gender in the mission-interested state-church milieus in Norway.

Individual opportunity and transnational education

The existence of several Norwegian mission societies implied greater competition for funding between the societies. For individuals, however, the existence of more than one Norwegian mission society provided greater opportunity for the realisation of a missionary calling. As long as the NMS was the main mission organisation in Norway, only those few men found fit according to the NMS' own criteria and capacity for education were accepted as missionary candidates at the mission school in Stavanger. In 1887 fifty candidates applied to NMS for the fifteen available places. Rejection was difficult to accept for an individual convinced of having a divine calling to become a missionary, and also for mission-friends and supporters of particular individuals (Kjebekk 1991). Among the first Norwegian missionaries sent to China by the NLCM, several had applied to the NMS mission school in Stavanger, but been turned down (Kjebekk 1991). Ole Næstegaard was another missionary applicant who had been turned down by the school in Stavanger.

There was no missionary education tailored for women in Norway until 1901 (Norseth 2007, Predelli 2003). In the NMS, only women who had already acquired an education were accepted as missionaries – or rather – as female missionary workers according to the society's use of nomenclature. Most of the women were teachers, some were nurses, midwives or deaconesses and a small handful was medical doctors. Only women from the middle and upper classes had the means to acquire higher education and thus were in a position to become missionaries. The missionary wives largely had the same background as the unmarried female missionaries (Predelli 2003).

In the NLCM too, the general rule in the first years seems to have been that women were already educated as teachers or nurses before being accepted as missionaries. With time, however, there seems to have been several examples of individuals merely having Bible school and/or shorter courses in nursing rather than full professional education.[70] What constituted a proper missionary-education was a contested issue within the NLCM. Allegedly, some influential mission-supporters were highly sceptical to a too long education for missionaries as it was feared this would damage the candidates' religious fervour (Kjebekk 1991). To what extent this scepticism was also a matter of class, not merely of religious ideology needs to be further examined. Because a formal education is a transferable social capital which may provide

[70] I base this on a review of the information on the missionaries provided in Vågen (1958). The prominent NLCM missionary and teacher Marie Monsen, was allegedly critical to this, arguing that it was necessary to have professional skills, not merely and primarily religious education in order to handle the requirements of missionary work in China (Golf 2000).

freedom and independence for the individual, it is necessary to try to understand why this is not considered attractive in certain religious milieus.

Since the NLCM initially did not have its own mission school, their missionaries-to-be had to be educated at other societies' mission schools outside of Norway. The missionary biographies show the extent to which the process of becoming a missionary implied entering a transnational and trans-denominational field of missionary education. The first eight missionaries sent to China for the NLCM in 1891, were the previously mentioned Johannes Brandtzæg (1861-1931), Ludvig Johnsen (1867-1893), Gertine Aarrestad (1859-1946), Ole Mikkelsen Sama (1864-1928), Henrik Seyffarth (1868-1933), Knut Stokke, Johan Albert Skordal (1890-1896) and Brita Vestervig (1867-1953).[71] While Johannes Brantzæg had a theological degree, Ludvig Johnsen had completed secondary school and had worked as a salesman before he became a missionary. Gertine Aarrestad was educated a deaconess in Oslo, but went a few months to London to prepare for her missionary work in China. Ole M. Sama had attended Bible school in Örebro, Sweden and an emissary-school in Bergen, before going to Red Wing Seminary in Red Wing, Minnesota. Also Johan A. Skordal, Brita Vestervig and Henrik Seyffarth were educated at Red Wing, the educational center for the Hauge's Norwegian Evangelical Lutheran Synod in America.

Another of the NLCM missionaries, Johan Andreas Olsen Gotteberg (1872-1955) first went to Britain in 1891-1893 attending courses at Harley College and Cliff College. Cliff College was a Methodist training centre for lay-people. Harley College was also called East London Missionary Training Institute. Founded by the famous Irish evangelist and preacher Henry Grattan Guinness and his wife Fanny, this college was a trans-denominational training centre for missionaries, predominantly for faith-missions. After Gotteberg had applied to- and was accepted as a missionary by the NLCM, he was sent to the Hauge synod's Lutheran Seminary in Red Wing for further education. He finished as a theological candidate in 1896.

There are several examples of how the mission-connection provided educational opportunities that otherwise would have been difficult to achieve. Ragnhild Botner (1868-1955) is one such example. The American-Norwegian Mission Society sponsored her medical education in the US. From Kristiansund, Botner was educated as a governess and had completed secondary education, before going to England in 1891. In 1898, she graduated as a medical doctor from the Women's Medical College in New York. She became a licensed practitioner of New York State University the

[71] This and the following biographical information from *Misjonsalbumet*, Gry forlag, Oslo (Vågen 1958) unless stated otherwise.

same year and went to China as a medical missionary for the American-Norwegian Society in 1898, working as a doctor in Hupeh.

The transnational education exposed the NLCM missionaries to other types of Christianity than the Norwegian Lutheran denomination. For instance, the Hauge synod stressed the significance of personal faith, and placed less weight on formal worship and rituals, corresponding to what can be termed a low-church tradition within Lutheranism and Protestantism. Gertine Aarrestad described the time she spent in London, '...alone among an unknown people, as some of the most educative experiences of my life' (*Kineseren* no. 2, 1892). She allegedly experienced richly God's faithful assistance and aid and through it, was confirmed in her calling to become a missionary among the Chinese. Aarrestad did not find the differences in faith and doctrine disturbing. However, there were examples of missionaries who subsequently struggled with their Lutheranism after the stay in London.

Once the NLCM had established its own missionary school in Norway in 1898, the educational connection to the US became less prominent.[72] Instead, many of the male missionaries attended six- or three-month courses at Livingston College in London, probably both to learn English and to get some basic instruction in medicine.

Boundaries, limits and risks

Johan Gotteberg and Ragnhild Botner got married in China in 1899. Their marriage illustrates another prominent aspect of the transnational missionary community; the extent to which individuals found partners in the field. Some found spouses within the same mission-society, such as Gertine Aarrestad and Ludvig Johnsen. Others married across denominations, mission-societies and nationality. Missionary in-marriage generally seems to have been unproblematic, but there were some clear exceptions to this informal rule. Firstly, the racial divide between Europeans or North-Americans and the Chinese was a significant boundary. When the first two Norwegian CIM missionaries Sofie Reuter and Anna Jakobsen both married in China, it was unproblematic that Reuter married one of the so-called Cambridge Seven, Stanley Smith (1861-1931). When Anna Jakobsen married the Chinese evangelist and former opium-addict Cheng Xiuqi, however, Hudson Taylor did everything he could to prevent it. When the wedding eventually did take place, it

[72] At first this school was placed in a small village in the countryside, some 80 km from Bergen (Framnes in Norheimsund). Then the school was first moved to Bergen and finally to Fjellhaug in Oslo, where it still is.

was considered a great scandal and Jakobsen Cheng was dispelled from the CIM.[73] It is evident from Taylor's own letters to Anna Jakobsen Cheng, that he thought this marriage across the racial divide endangered the entire mission.

In his opinion, Anna Jakobsen set all Western single female missionaries in China in danger through her action. While the Christian mission-project implied the transgression of national, cultural and religious boundaries, this case illustrates how the justification of such a movement in part rested on an underlying non-negotiable boundary between specific notions of the relationship between 'us' and 'them'. In most mission-contexts at this point in time, and also in most colonial contexts, inter-racial marriages were considered problematic and difficult to handle, however, more difficult in the instances when European women married local men than vice versa. Interestingly, the fact that they were expelled from the CIM, did not – at least in a very visible way – change Anna Cheng and Cheng Xiuqi's situation or living. They merely started their own mission, called Anna Cheng's Mission, thus illustrating the individual freedom that faith-missions could imply in instances where the individual missionary clashed with the mission-leadership. The fact that Anna Cheng's mission retained the financial support from the free-church milieus in her hometown, underlines the special and often close relationship between individual missionaries and mission support groups which such missions sometimes fostered. These China-mission supporters even continued to support Anna Cheng's mission after Anna Cheng's premature death in 1913.

Another limit regarding marriage across mission society boundaries was connected to the relationship between mission-societies' financial investment in missionary candidates, and these candidates' length of active service in the mission-field. When Ragnhild Botner and Johan Gotteberg soon upon their marriage received a letter and a claim for a monetary refund of everything which Botner had cost the American-Norwegian mission-society, this was probably due to a combination of her lengthy education and the fact that she had only been an active missionary in China in one year before she got married. The wife tended to marry into the husband's mission society. So too with Ragnhild Botner. The claim from the American mission society was sent to the NLCM, since the Gottebergs were in no position to afford such a large sum. The American-Norwegian Society refused to agree to down-payments. Instead it demanded that the sum was paid immediately and in its entirety: Botner's medical education (US$ 1,300), her journey to China, her travels for the mission in the US,

[73] Both Reuter, Stanley and Xiqui worked in close cooperation with the Chines Protestant pastor Xi Shengmo (Pastor His). Renowned for his treatment of opium slaves, Shengmo was the founder and leader of one of the largest indigenous Chinese Protestant sects between the Taiping- and Boxer rebellions (Austin 2007).

and even her wage for the first year in China.[74] In a letter to the NLCM board, Gotteberg wondered whether this was a tactical move either to get as much money as possible from the NLCM, or an effort to intimidate the couple into the American-Norwegian Society's service. They were told that the American-Norwegian Society would withdraw their charges provided they became their missionaries. Also the NLCM requested refund when their female missionary Brita Botnen got engaged to be married with the pastor Ivan Ehne from the Swedish *Svenska Missionsförbundet.* This was apparently done according to established inter-mission rules. In both cases, it was alluded to the fact that it was in accordance with Chinese customs of conduct that husbands had to pay for their bride.[75]

This illustrates how missionaries in a sense were commodities in this large transnational 'world wide web'. It also shows how investment in missionaries was a risk. The mission-friends' and mission societies underlying investment premise was trust; an explicit or implicit agreement between the parties that the individual would work long enough in the mission for the investment to be worthwhile. Yet, there were many circumstances that could not be controlled, the main being illness and death. Ragnhild Gotteberg's example also indicates why investment in women could be more of a risk than the investment in men. In another mission-context, the likelihood that single female missionaries would marry before long was used to argue why the mission should not take the extra cost of constructing separate dwellings for single female missionaries (Skeie 2012). Spiritual disagreement was another risk.

In 1901 after ten years' hard pioneer work in China, following the devastating material damages of the Boxer rebellion, the NLCM lost six of its most well-educated missionaries. Ragnhild and Johan Gotteberg, together with Arthur (1870-1941) and Fredrikke Hertzberg (1873-1930), Nils Arnetvedt (1846-1935) and the missionary candidate and medical doctor Jørgen Nilsen (1871-1922) all resigned due to spiritual disagreements with the NLCM regarding the missionary education and role. Following his theological education at the Luther Seminary of the Hauge synod in Minnesota, Gotteberg wanted to be ordained as a Lutheran pastor. By this time, the NLCM had taken a principal ideological stand against pastoral ordination of its missionary candidates. Gotteberg's request was therefore turned down. Arnetvedt had been ordained in the US, because the NLCM leadership had been unable to prevent it. Hertzberg and Nilsen supported Gotteberg and Arnetvedt's theological views in the matter, and the six felt compelled to resign.

[74] Letter from J. Gotteberg to NLCM board, Tsin-shan-Kiang 2. January 1900 in NLM Archives Main Office, Box DB 0003.

[75] Letter from Laohokow, 22. Jan. 1932 in NLM Archives, China and Hong Kong Field-archives, DA-0004, folder 1.

As it turned out, they were not idle for long. In 1901 the NMS General Assembly had decided to take up missionary work in China after all. Now all the resigned NLCM missionaries became NMS' first representatives in China. The ideological reasons for these missionaries' resignation were real enough, and had been made thoroughly visible through a formal meeting between the missionaries and the NLCM leadership. However, the already tense relationship between the NMS and the NLCM deteriorated further as the NMS took over their missionaries so swiftly and got a flying start in China harvesting NLCM's missionary investment.

The conglomerate of mission societies working in China and the fact that (able) missionaries at least to some extent were a commodity in demand, implied that individuals could and in some instances did manipulate and move between mission societies according to their own interests. Missionaries who had difficulties cooperating with others, and individuals with psychological issues could cause great damage in mission societies and for individual missionaries. As if it was not enough to lose many of its most able missionaries, the NLCM also had its share of missionaries unfit for missionary work during this first period. Sofie Clausen, who arrived in 1892, was diagnosed as sick and forced to go back to Norway in 1896. Lyder and Inger Kristensen were a couple which had begun as missionaries for the Kristiania support-group of the NLCM. Before he returned as an NLCM missionary, Lyder Kristensen had served in the American Simpson's Alliance mission. Already on the ship taking them back to China, the difficulties started for their companion, the young single female missionary Marie Monsen (1878-1962). All major decisions in the field had to be approved by the mission board in Norway, and despite the telegraph being introduced in China in the 1870s – correspondence between China and Norway still took its time. Hence, things that had begun to go wrong could not quickly be set straight. The problems caused by the Kristensens, allegedly made Monsen ill and on the brink of giving up altogether (Golf 2000). What finally brought the problems into the open, and forced the NLCM to sack the couple, were grave complaints from some Chinese co-workers.[76] Before long the two were working for the United Church mission another place in China. For a number of missionaries, the challenges in China with recurrent violent uprisings and strong anti-foreign sentiments, was psychologically demanding. For some it was too much; Ole Næstegaard, who had shown previous tendencies of having an unstable psyche, got so ill following the Boxer-rebellion, that he had to be sent back home from China.

[76] NLM Archives, Main Office, Dea 0002, Conference minutes 1907.

Together these cases illustrate crucial aspects of the underlying mutual trust that the missionary network thoroughly depended on to function properly. The mission societies did their best to minimise and control risk. The same was true for the NLCM, who decided that creating their own missionary education was an important investment in several mutually dependent factors. Not exposing young individuals to educational influence from other denominations reduced the risk of the alienation of missionary candidates to the society's particular understanding of Christianity. Instead, their own missionary education gave a unique opportunity to educate individuals in faith's fundamental truths and the missionary work's basic foundations, according to NLCM's theological and missiological position. Monitoring the candidates over time, also gave an opportunity for a more thorough evaluation of the candidates' personality and skills to become missionaries. Finally, education in Norway also kept the missionary candidates closer to the mission-friends and supporters in Norway, which in turn made fund-raising easier than it was when the candidates spent so much time on a school far away in the USA.

Transnational mission networks in China

Benedict Anderson's notion of 'imagined communities' (1983) is helpful in order to understand one aspect of the power of the transnational missionary network. Just as individuals became convinced and inspired by reading of and listening to tales of the works and wonders of prominent missionaries like Hudson Taylor, it was also significant to narrate individual missionary lives according to the same formula. Printing the Norwegian missionaries Gertine Aarrestad, Johannes Brantzæg, Ole M. Sama and all the other missionaries' life-accounts next to extracts from the great China-missionary Hudson Taylor in *Kineseren* [The Chinese], the mission magazine literally and symbolically made the Norwegian missionaries and mission part of the greater Protestant mission cause in China and the world. Being sent to study English among many nationalities and missionaries-to-be in London, underlined this transnational community of Protestant missions as fundamentally working for the same cause.

This was even more evident as the missionaries arrived in China and used the missionary network for everything from housing to recruiting Chinese evangelists. Missionaries from other Scandinavian countries and Scandinavian-American societies constituted one significant informal network for the Norwegian missionaries in China. As the first missionaries from the NLCM landed in Shanghai, they were welcomed by Norwegians – the independent missionary Ludvig Eriksen whom Brantzæg knew from before, and Sigvald Netland who had initiated the establishing of The Norwegian Lutheran China-mission Association in America.

Allegedly, they found Shanghai so westernized that they longed to get to know 'the real China' (Tiltnes & Vig 1946). Not before long they settled down in Hankow and Wuchang, where Griffith John of the LMS had worked since 1861. The Swedish Missionsforbundet and an American mission, later named Hauge's synod, was also working in Wuchang. The NLCM missionaries used the first year to learn Chinese and to look around for a possible and suitable mission field. In 1892, they started up in Laohokow, a place where CIM had a station. At the same time they built a 'mission-home' in Hankow, which housed not merely the Norwegians but also the first missionaries from the Danish mission association pastor Nyholm and his wife and another female missionary, Caroline Johansen. Later, mission-stations were established in Fangshien, Shihwarai, Künchow, Yünyang, Sichwan, Tengchow, Chenping, Nanyang and Lushan. From these main-stations, missionary work was carried out in the surrounding district aided by Chinese workers, that is, male evangelists and female Bible-women. During the first years, the NLCM depended on the CIM network for their Chinese workers. Subsequently, they started their own schools for educating Chinese workers in the mission (Tiltnes & Vig 1946).

The extent to which the missionary network also implied the promotion of certain conduct and particular strategies for missionary work is evident. Not least in the importance placed on wearing Chinese clothes. In the CIM, to wear Chinese clothes and even braid their hair in a long 'pig-tale' was the result of a difficult decision to 'go native' (Austin 2007: 2, 121). While the CIM policy on Chinese dress may have been more extreme than others, many missionaries besides CIM missionaries wore Chinese clothes at different time-periods and in different contexts, particularly in rural, anti-foreign contexts. From a historical point of view, the CIM continued a custom which Catholic Franciscan missionaries had begun in the18th century and kept up until the 1940s (Harrison 2013: 120). Also the German Protestant missionary Gützlaff (1803-1851) is known for having used Chinese clothes (Lutz 2008). The mere thought of dressing up in order to become more similar to the local population was unthinkable in most other mission contexts.

Conclusion

The Norwegian China missions operated in a large transnational space characterised by mutual informal cooperation and contact both in Europe and in China. While the period was characterised by a great deal of trans-denominational and transnational mission-cooperation internationally, this informal contact was within a few years after NLCM's founding more prominent for the missionaries in China than it was for the NLCM in Norway. The primary reason for this was the development towards religious and organisational differentiation in Norway, and the NLCM's role in the self-conscious lay-groups' development of an institutional platform independently from the Norwegian Lutheran state church. While the individuals in the NLCM were for the most part members of the state-church, the importance for these lay-groups was to not be under church-leadership, underlined in the slogan '*In* the church and *with* the church, but never under church leadership' (Seland & Aagedal 2008: 36; my italics and translation). The NLCM was in an important sense a materialisation of this slogan.

The CIM and Hudson Taylor were directly and indirectly significant for the establishing of all the Norwegian China-missions, however, only the free-church milieus cooperated formally with the CIM and took the example from the faith-mission-organisation of the work in Norway and in China. For the NLCM it was important to mark distance from the CIM to gain legitimacy and respectability among its supporters. While the NMS was a traditional board-mission also in China, the NLCM interestingly represents something in-between. The NLCM too was a board-mission, but organised the work in the field and the relationship between the missionaries in the field and the mission-supporters in Norway in ways that reflected faith-missions. Both the NMS and the NLCM's informal practical use of CIM networks and Chinese workers with CIM background in China was contrasted with a denominational Lutheran position in opposition to the CIM's non-denominationalism. Thus the case of the Norwegian China missions also reflects the various ways transnational impulses were appropriated and transformed locally, both by the missionaries in China and by the religious milieus in Norway.

Acknowledgements

The author would like to thank the Norwegian Research Council for funding of the work presented here through Project Number 205553.

Bibliography

Austin, A. 2007. *China's Millions. The China Inland Mission and Late Quing Society, 1832-1905*. Studies in the History of Christian Missions, William B. Eerdmans Publishing Company, Grand Rapids, Michigan, MI, USA/Cambridge, UK.

Bays, D. H. 2012. *A New History of Christianity in China*. Wiley-Blackwell Publishing, West Sussex, UK.

Brandtzæg, J. 1902. Hvordan gik det til at Kina-missionen begyndte i Norge. Særtrykk av *Kineseren* for 1900. Det Norske lutherske Kinamisjonsforbund, Bergen, Norway.

Bundy, D. 2009. *Visions of Apostolic Mission. Scandinavian Pentecostal Mission to 1935*. Acta Universitatis Upsaliensis: Studia Historico-Ecclesiastica Upsaliensia 45, Uppsala, Sweden.

Claesson, A.M. 2001. *Kinesernas vänner: En analys av missionens berättelse som ideologi och utopi*. PhD Dissertation Lund University, Jönköping läns museum, Jönköping, Sweden.

Cresswell, T. 2006. *On the Move: Mobility in the Modern Western World*. Routledge, New York, NY, USA.

Golf, O. 2000. *Marie Monsen: Kinamisjonær, bønnekjempe, kvinneaktivist*. Lunde, Oslo, Norway.

Gullestad, M. 2007. *Picturing Pity: Pitfalls and Pleasures in Cross-Cultural communication: Image and Word in a North Cameroon Mission*. Berghahn, New York, NY, USA.

Hannerz, U. 1996. *Transnational Connections: Culture, People, Places*. Routledge, New York, NY, USA.

Harrison, H. 2013. *The Missionary's Curse and Other Tales from a Chinese Catholic Village*. University of California Press, California, CA, USA.

Hofmeyr, I. 2005. Inventing the World: Transnationalism, Transmission, and Christian Textualities. In: Scott, J.S. & Griffiths, F. (eds.). *Mixed Messages: Materiality, Textuality, Missions*. Palgrave Macmillan, New York, NY, USA.

Iriye, A. & Saunier, P.-Y. 2009. Introduction. In: Iriye, A. & Saunier, P.-Y. (eds.). *The Palgrave Dictionary of Transnational History: From the mid-19th century to the present day*. Palgrave Macmillan Transnational History Series, Hampshire, UK.

Kineseren No. 2, 1892.

Kjebekk, E. 1991. *Verden for Kristus: Norsk luthersk misjonssamband 1891-1991: heimearbeidet*. Lunde, Oslo, Norway.

Lutz, J.G. 2008. *Opening China. Karl F. A. Gützlaff and Sino-Western Relations, 1827-1852*. Studies in the History of Christian Missions, William B. Eerdmans Publishing Company, Grand Rapids, Michigan, MI, USA/Cambridge, UK.

Mikaelsson, L. 2003. *Kallets ekko. Studier i misjon og selvbiografi*. Høyskoleforlaget, Bergen, Norway.

NLM Archives, China and Hong Kong Field-archives, DA-0004, folder 1, Letter from NLMS superintendent Laohokow, 22. Jan. 1932

NLM Archives, Main Office, Dea 0002, Conference minutes 1907

NLM Archives Main Office, Box DB 0003, Letter from J. Gotteberg to NLCM board, Tsin-shan-Kiang 2. January 1900

Nome, J. 1943. *Det Norske Misjonsselskaps historie i norsk kirkeliv*. Dreyers Grafiske anstalt, Stavanger, Norway.

Norseth, K. 2007. *'La os bryte over tvert med vor stumhet!' Kvinners vei til myndighet I de kristelige organisasjonene 1842-1912*. Det teologiske menighetsfakultet, Oslo, Norway.

Nielssen, H., Okkenhaug, I.M. & Skeie, K.H. 2011. Introduction. In: Nielssen, H., Okkenhaug, I.M. & Skeie, K.H. (eds.). *Protestant Missions and, Local Encounters in the Nineteenth and Twentieth Centuries. Unto the Ends of the World*. Studies in Christian Mission vol. 40. Brill, Leiden, the Netherlands.

Predelli, L.N. 2003. *Issues of Gender, Race, and Class in the Norwegian Missionary Society in Nineteenth-Century Norway and Madagascar.* Studies in history of missions, Edwin Mellen Press Ltd, New York, NY, USA

Reinders, E. 2004. *Borrowed Gods and Foreign Bodies: Christian Missionaries Imagine Chinese Religion.* University of California Press, Berkeley, CA, USA.

Seland, B. & Aagedal, O. 2008. *Vekkelsesvind. Den norske vekkingskristendommen.* Det norske samlaget, Oslo, Norway.

Skeie, K.H. 2012. *Building God's Kingdom. Norwegian Missionaries in Highland Madagascar 1866-1903.* Studies in Christian Mission, vol. 42. Brill, Leiden, the Netherlands.

Stanley, B. 2009. *The World Missionary Conference, Edinburgh 1910.* Studies in the History of Christian Missions. William B. Eerdmans Publishing Company, Grand Rapids, Michigan, MI, USA/Cambridge, UK.

Tiltnes, A. & Vig, A. 1946. *Det norske lutherske kinamisjonsforbund gjennom 50 år.* vol. II, Det norske lutherske kinamisjonsforbunds forlag, Oslo, Norway.

Tyrell, I. 2010. *Reforming the World. The Creation of America's Moral Empire.* Princeton University Press, Princeton, New Jersey, NJ, USA.

Vågen, R. 1958. *Misjonsalbumet: Norsk Luthersk Misjonssambands misjonærer og misjonsmarker 1891-1957.* Gry forlag, Oslo, Norway.

Von Bülow, M. 2010. *Transnational Networks. Civil Society and the Politics of Trade in the Americas.* Cambridge University Press, Cambridge, UK.

Chapter 5: Regional resonances

Vietnamese whale worship in comparative perspective

Michael Parnwell

> *Cẩm Thanh Fisherman*: 'There are still fishermen who are helped by the whale, rescued at sea.[77] Something like this happened here around two months ago.[78] There was a big boat, like a fishing boat, that was carrying 35 soldiers from the Cham Islands to their homes.[79] But, that day the weather was not good, and when the boat was still in the ocean, one of the soldiers – he didn't live on the island; he lived on the mainland, so he never thought about the whale, he never knew about it – saw something behind the boat, like a fish, but he didn't believe that it was a fish. He asked the boat captain "why are we returning to the shore with something behind us?" The boat captain told him to keep silent. Yes, just to keep silent. When they came close to the mainland a big wave came suddenly and the boat capsized. A lot of soldiers were saved, but four (in fact seven) of them were killed.[80] He said that the whale followed the boat to the mainland, but when the whale finished its journey, that's when the accident happened.'
>
> *Me*: 'Why did the boatman say "shhh, be quiet"?'
>
> *Fisherman*: 'Because, the whale is considered to be a rescuer, a saviour. When a whale dies, it is considered like a god. So that means that they respect the fish. But, if they talk about it, that means that maybe they cannot get help.'
>
> *Me*: 'That's what I heard. You must not use the name "whale", you must call it "Mr Fish" or "Lord Fish".'
>
> *Guide*: 'That's why we call it "Ca Ông" – that means Lord Fish. This man is the age of my grandfather, but he still says "Ông", that means a thing that is respected.'[81]

[77] Interview with a retired fisherman, Cẩm Thanh village, Quảng Nam Province, Việt Nam, 4 April 2012.

[78] 25 December 2011.

[79] The boat was actually carrying 28 soldiers from the Army base on Cù Lao Chàm, and six residents of Tân Hiệp Commune, in addition to the boat captain.

[80] In fact, seven people drowned in this accident, five soldiers and two residents of Tân Hiệp Commune. Available at: http://english.vietnamnet.vn/en/society/17180/bodies-of-five-soldiers-from-capsized-boat-found.html, although local sources put the death toll at 13.

[81] The proper title is in fact *Cá Voi* [elephant fish], but it is more commonly referred to as *Ca Ông* [Grandfather Fish] if the whale is male, and *Bà Ngu* [Grandmother Fish] if female (Kiem 1996). A whale that has drifted ashore will be called *Ông Lui* (Lantz 2009).

C. Brautaset and C.M. Dent (eds.) **THE GREAT DIVERSITY**
DOI 10.3920/978-90-8686-786-8_5, © Wageningen Academic Publishers 2014

> *Me*: 'So, they thought that the whale was protecting the boat from the very bad weather, but then later the whale went away and then the accident happened?'
> *Fisherman*: 'If you keep silent, that means that you believe in the whale. If you talk about it, that means that you don't believe it, you don't show respect to it.'
> *Me*: 'So, the soldiers didn't understand, but the boatman, he understood?'
> *Fisherman*: 'He understood the story very well.'
> *Me*: 'Is it possible that somebody did speak, and that this is why the protection was lost, and then they died, possibly?'
> *Fisherman*: 'Possibly. But later, the four (five) soldiers who died in the ocean, later the whale took each of them to the mainland, but to different places on the coast, exactly to where they lived. For example, if he was living in Da Nang, the body drifted towards Da Nang. Another one who lived in the south, the body drifted in that direction; a few days later after he died, the body was taken to the other side, close to his family.'

The story above is a slightly unusual variant of the 'rescue narratives' that are quite commonly heard from fishers along the coast of central-south Việt Nam. I have frequently listened to older members of fishing communities telling similar stories of incidents when fellow villagers have been caught by storms while out at sea, or their craft and thus their safety has been threatened in some way or another, only for a whale to appear close to their boat, either calming the waves or helping to take the boat closer to shore where they would be safe. Partly because of the protection that whales are believed to provide fishers and other travellers while at sea, and partly because of fishers' belief that the whale is a distant ancestor and a physical manifestation of the deities that inhabit the marine environment, a complex series of veneration practices has evolved which include the organisation of funerals for deceased cetaceans, the construction of whale temples, and the propitiation of the spirits of the sea.

Although whale veneration is a quite unusual and distinctive characteristic of coastal fishing communities in central and southern Việt Nam, stretching from Da Nang to beyond the Mekong delta (Lantz 2009, Macdonald 2003), in essence it is not unique to the country, either in time and space. For instance, picking up the themes of nomenclature and silence from the narrative above (see also Kiem 1996, Macdonald 2003), we can find similar instances from other parts of the Asia-Pacific region, and further afield. Nelly Naumann, writing in 1974 on Japanese whale cults, noted that while whales and sharks were said to help fishers by chasing shoals of fish into their nets if they were treated reverentially, the opposite would be the case if the sea mammals and the spirits they harboured were angered by loud noises or inappropriate behaviour (Naumann 1974). After a successful whale hunt among the

Koryaks of the Russian Far East, the community entered a period of mourning for the soul of the deceased whale, during which time '...no loud noise [was] allowed to disturb him, children [were] not allowed to cry' (Lantis 1938: 446). During a Makah whale hunt off the coast of Washington state in the United States '...everyone would try to keep quiet and move gently' (Miller 2000: 186).

Even the hunting of sea turtles in Fiji requires members of the community to remain quiet, both in the village and on the shore, so as not to displease the turtle's spirit: the belief – which also has resonances elsewhere in the region, as we shall see shortly – is that the spirit decides to deliver its body for the benefit of the community (Morgan 2007), but only if treated with respect and decorum. Aboriginal Saltwater People in Australia used to 'sing [ritually call up] sharks and dolphins to drive fish towards the people on the shoreline where they surrounded the fish and picked them up' (McNiven 2004: 339). On the Indonesian island of Lamalera, studied by Barnes (1974) in the 1970s, the traditional whalers would not converse, or would only talk in a subdued manner, during the course of a whale hunt until a whale had been killed. If whalers' children were rowdy at home while a hunt was taking place, the hunt would be unsuccessful (Barnes 2005).

Also, when encountering a whale, the whale-hunters must not use their proper names, using only the generic *bapa* (a word of respect used to address a senior male), nor must the hunters use the formal name of the whale, *kotan kelema* (sperm whale, in this instance), using instead a generic *ikan* or 'the fish', a practice that is also found in other parts of Indonesia although not specifically in connection with whales (Barnes 1974). In Nusa Tenggara Timur, the whale shark is believed to be an 'ancestor of the sea' [*mbo madilao*] of the Bajau and a manifestation of one of the spirits of the sea, so accordingly it is forbidden to use its formal name in everyday conversations (Stacey *et al.* 2008).[82] Spirits of the sea were prone to anger if sailors '...ignored prescribed ritual or employed inappropriate words, especially those associated with the land' (Andaya 2006: 680).

The aim of this chapter is to contribute to the illumination of 'The Great Diversity' of the Asia-Pacific region by introducing the practice of whale veneration in Việt Nam and undertaking a preliminary exploration of what I call 'regional resonances' in traditional societies' relationship with the animals – particularly sea mammals, and especially cetaceans – and deities that are believed to inhabit the region's seas and oceans. The relationship between people and whales has a chequered character:

[82] There is also a 'rescue narrative' of Bajau fishermen being saved from a sinking boat by a whale shark, which allowed them to cling to its dorsal fin as it took them to safety (Stacey *et al.* 2008).

in some places whales are worshiped, respected, admired, watched and protected; elsewhere, they are hunted, slaughtered, consumed, traded and feared, and have even become a source of international political tension. The following discussion shows that, even across this exploitation/conservation divide, there are multiple common manifestations of people's spiritual and physical interactions with the sea that are suggestive of a more sensitive array of attitudes, beliefs and practices than is sometimes evident from the whaling debate.

The following discussion will outline some of the principal features of whale worship in Việt Nam[83], and in each instance explore resonances, variances and dissonances across the diverse cultural and geographical spaces of the Asia-Pacific region (and occasionally a little further afield). Because of constraints of space, this preliminary analysis will only focus on two key themes: manifestations and motivations. To my knowledge this is the first paper to attempt to look at whale veneration cross-regionally. Charles Macdonald (2003: 132) raises the question whether there are cults to marine mammals elsewhere in Southeast Asia that are similar to whale worship in Việt Nam, but restricts his investigation mainly to some of the islands of the Indonesian archipelago, and comes up with the conclusion that the whale cult is '*une exception Vietnamienne*'. Is this the case?

Rites for whales

In central-south Việt Nam, if a dead whale is found beached or drifting close to the shore, the nearest community will organise a funeral, either within the village if the carcass is small enough to be carried, or *in situ* if not. The discovery of a dead whale will typically be seen as an omen of good luck for the community, and especially the person who discovers the body (Huy *et al.* 2003). The funeral rites will be largely the same as for a local person of high standing (Kiem 1996). In other words, the whale will be afforded human status because it is seen as an ancestor of fishermen. Fishers from the community will not go out to sea for a month following the discovery of the whale's body, out of respect (Lantz 2009). No seafood will be available during the funeral festivities, as the whale is considered to be the vehicle for the god of all sea creatures, and no necklaces made of claws will be worn as they are believed to make the whale's spirit feel uneasy. These and other prohibitions are keenly respected, as some fishermen believe that if they offend the spirit of the whale in any way it will not afford them any protection when they are out at sea.

[83] Unattributed information used in this discussion is derived from my own research of 20 whale temples in the area between Da Nang and Hội An, in Quảng Nam Province, in April 2011 and April 2012.

Fishers from neighbouring communities[84] will also sometimes be invited to the funeral (see also Macdonald 2003), in part because veneration of the dead whale's spirit is observed by the population of fishers as a whole, but partly also in order to help defray the costs of holding the funeral.[85] The person who discovers the dead whale will enter a period of mourning (usually for three years but in some communities I visited this was one year or even three months), in exactly the same manner as would apply to a senior member of their own family.[86] He will be considered a 'son' of the deceased whale, through which the latter will communicate its wishes with regard to the burial (Macdonald 2003). In one community near Hội An, Duy Xuyen, lying just south of the Thu Bồn river, I was told that a person in mourning would not, for instance, walk under a washing line as this would require him[87] to bow his head, and whilst in mourning he should only bow to Lord/Sir Fish [*Cá Ông*]. The whale will be interred for a period of three years (in some communities this was just one year),[88] before being disinterred,[89] the bones ritually cleaned and blessed in a ceremony that Macdonald (2003) described as a 'second funeral', wrapped in red cloth and votive paper and then placed in coffins or a vault within the local whale temple (Kiem 1996). If one does not already exist, a new whale temple may be established when a whale funeral is held. Inside the whale temple there will typically be an altar (the central one) to the whale genie, flanked by altars to other deities of the sea. Each year the community will hold at least one 'memorial day' ceremony, either marking the date of a particular whale funeral or as a date to commemorate all deceased whales and to pay obeisance to the whale genie.

Whilst there are few specific international examples of whale funerals along lines identical to those described above for Việt Nam, we can find many similarities cross-regionally in both function and form. Some of these will be briefly outlined

[84] Sometimes, usually when a dead whale is particularly large, and its discovery gains publicity in newspapers or on television, many hundreds or even thousands of people will come to look at the dead creature, and as many as 10,000 people may attend the deceased whale's funeral (*Huffington Post*, 23 February 2010).

[85] The cost of the funeral may be in the region of US$2-3,000, according to a respondent from Hà My village (personal communication, 5 April 2011).

[86] The whale is viewed as the ancestor of fishers, and thus whale veneration is one of many forms of ancestor worship practiced in Việt Nam where people believe that a deceased family member has a continued existence and continues to take an interest and have some involvement in the well-being of those left behind (Lantz 2009).

[87] In all instances that we were told concerning the discovery of whales and entering a period of mourning the subjects were all males, reflecting the fact that fishing is almost exclusively and occupation for men. Women, however, are actively involved in the memorial day celebrations.

[88] In one community, reported by Smith *et al.* (1997), the whale bones were cremated and the ashes were placed in coffins within the whale temple. This community was not actually located on the coast, as is more typically the case, but some 40 km upstream along the Mekong River.

[89] The community in Binh Minh neither disinters the whale nor store the cleaned bones in coffins within the whale temple – illustrating how varied whale veneration practices are in Việt Nam.

here. Naumann (1974) describes cemeteries in the former local whaling districts of Japan which contain memorial stones that are dedicated to whales, the inscriptions on some of which implore the rebirth of the spirit of the whale as a Buddha. It is believed that Ebisu-gami, the principal deity of fishers, may take the form of a whale or a shark, and is one of seven Deities of Good Fortune. If the community killed more than 35 whales it was obliged to construct a memorial grave containing some of the bones of the whales, and organise masses or ceremonies for the spirits of the deceased cetaceans. Shindo (1975) describes graves for whales in Fukuoka Prefecture which were intended to calm the spirits of the dead whales, and also (1975) refers to Kongoji Temple, in Akihama-cho, which is dedicated to whales, and where rites to whales have been performed since 1837. It is understood that the practice commenced in 1806 upon the capture of a whale in a nearby creek, the captors being surprised to find that the whale's blood was warm. Segi (2003), writing about Taiji, one of the original locations of traditional whaling in Japan, describes unique whale ceremonies and festivals linked to requests for successful whale hunts and good whale catches. Creason (2004) describes memorial ceremonies for the souls of killed whales, and death registers being kept with the names of dead whales, seeking '...forgiveness and spiritual compensation for the loss of karmic merit that results from the taking of life' (Creason 2004: 100).

On the island of Lamalera in eastern Indonesia, before the advent of Christianity in this location, there was a 'temple for the fish' [*Korke Ika*] or a 'house that contains the fish' [*lango ika belifu*] – in both instances fish is a pseudonym for the whale – where 'parts of the head of the sperm whale go to the lord of the land, so that he can store the soul of the animal in his temple' (Barnes 1996). The village has four entrances, one of which is called the *ika kota*, which is where the skulls of whales, rays and other large sea creatures are placed (Barnes 1974). Before the commencement of whaling each year, the lord of the land would conduct a number of rituals, including visiting a boulder in the shape of a whale, in order to ensure a successful and safe whale-hunting season. Not only would bones from whales be stored in 'fish temples', but so too would the skulls of ancestors be kept in the sheds where whaling boats were stored during the off-season in order to afford protection and prevent the encroachment of dangerous spirits. The whalers believe that the whale continues to have power even after it is dead, and that it is capable of identifying (physically and spiritually) weak points in their craft, so considerable efforts are made to ensure that boats are constructed in a ritually appropriate manner, tools are properly blessed before being put to use, and animal sacrifices are made to increase the prospects of a successful whale hunt (Barnes 1974). With the expansion of Christianity in this area, many of these traditional practices have disappeared, and now whale bones are cremated to prevent them from littering the beach (Barnes 1996). A similar situation was found in the Philippines where, in the traditional whaling community on Pamilacan Island in the Sulu Sea, whale numbers were thought to have declined because people had stopped making their traditional *buhat-buhat* offerings to the spirits of the sea before the commencement of the whaling season (Acebes 2009). With the spread of Islam, some communities will no longer consume whale meat because cetaceans are considered to be *baboy sa dagat*, or 'pigs of the sea', and thus *hukum* to Muslims.

Lantis (1938) described solemn rituals among the Koryak of the Russian Far East following the capture of a whale, intended in equal measure to honour the whale as well as its captors. These included placing a cup of fresh water before a wooden image of a whale on a shrine, taking the hump of a whale into the whaler's house where it was set in a place of honour, and the ritual treatment of bones intended to ensure that the whale's spirit could return to the oceans uninjured. A period of mourning lasting three days was also held in honour of the deceased whale's spirit, before the whale is offered food for its return journey to the sea.

Ritual burial mounds of dugong – another sea mammal, but a sirenian not a cetacean – have been discovered in the Torres Straits dating back to the 13th or 14th centuries (David *et al.* 2009) and are said to reflect a spiritual connection between hunter and prey, which become one after the death of the dugong (McNiven 2010). As with the case of Indonesia above, dugong hunters believed that the animal continued to have a spiritual existence beyond its physical death, and the intricate laying out of dugong bones – skulls in the middle, surrounded by rib bones, and occasionally also including ear bones to allow the hunters to utilise the dugong's sharp sense of hearing – was a reflection both of the magic Torres Straits hunters used to sharpen their hunting edge and of their spiritual connections to the world they inhabited (McNiven 2004): the remains of their prey, and their engagement with the spirits of these prey, were considered to be essential tools in their quest for more prey (David *et al.* 2009). Sea mammals thus form an integral part of traditional societies' 'spiritscapes' (McNiven 2004).

The motivation and context for these spiritual actions taken by societies which see whales and other sea mammals as prey was clearly quite different to the case of the Vietnamese, who never hunt or consume the body parts of whales, and yet there is a clear resonance in terms of respect for the animals and their spirits on either side of the exploitation/protection divide. The funeral rites in Việt Nam are a sign of respect for the spirit of the whale-as-ancestor, although as we shall see in the next section there may also be some utilitarian purpose for these spiritual practices. Whilst Vietnamese fishers anticipate that the whale genie will protect them during their dangerous voyages at sea, the Japanese and other traditional whalers wish to appease the spirit of the whale they have killed, to prevent it from being vindictive and dangerous, but also so that the suitably propitiated whale spirit will again put itself at the disposal of the community, arriving at some future date with another body which they can then utilise. This is a common motif among traditional coastal communities. The Makah of northwest USA consider whales to be gifts from the gods: a whale[90] is too large for a human to capture, thus the deliverance of any whale to them – even through their own hunting effort and prowess – must be because the animal wished to make itself available to humans for consumption (Miller 2000). As Sullivan (1998) comments, the whale chooses the whaler. In addition to consuming whale meat and using oil and other items, the Makah will use whale bones, teeth and other body parts for their artwork and for ritual adornments, which often have ceremonial uses. Makah whalers will prepare themselves spiritually, sometime for several months before the whaling season commences, retreating to the woods to practice their rituals, imitating whales while bathing and swimming, reciting particular powerful songs, and propitiating their animal guardian spirits. Once a whale had been struck, further songs and incantations would be uttered in order to encourage the whale to turn towards the shore and towards the community members who '…stood ready to give it praise … and to honour it as a guest of the village with ceremonies and rituals' (Miller 2000: 186).

According to the Māori who, like the Vietnamese (Smith *et al.* 1995), abhor the hunting of cetaceans, whales are equivalent to gods, which are highly respected and are believed to be deities which afford protection to Māoris who propitiate them. Whales are seen by the Māori of New Zealand as having guided and protected their canoes during their oceanic migrations. The Māori will never kill whales, but may utilise the products of the carcass of a beached whale (Creason 2004) which, as with the Makah, they will consider to be a gift to them from the gods.

Whales as compassionate, benevolent and spiritual beings

Why do people from different and seemingly quite diverse ethnic, social and cultural backgrounds venerate whales and other sea mammals, and treat them with respect, even when hunting? We will see that, whilst the motivations for whale worship are superficially quite different, ranging from the deeply spiritual to the ostensibly

[90] The Makah mainly hunt Grey Whales, *Eschrichtius robustus*, which may be 15 metres in length and 40 tonnes in weight.

utilitarian, there may also be some common roots and manifestations which resonate across space and time.[91]

It is a widely-held belief among fishers in central-south Việt Nam that whales will come to their rescue if they are in peril at sea, provided they believe in the whale-genie, act in an appropriate manner, and have maintained their devotions to the whale spirit or the spirits of the sea for which the whale is believed to be a conduit. Before looking at the origins and character of this belief in a little more detail, I would like briefly to reflect on the apparent incongruity between people's spiritual and utilitarian faith in the whale-genie, based on my own field-work in the area around Da Nang and Hội An. In a purist sense, whale worship should consist of the veneration of cetaceans and their spirits for its own sake, rather than for the benefits it yields. However, enquiries about any such tension were generally met with polite incredulity.[92] Ancestor worship in Việt Nam is based on the belief that a family member, when they have passed from the mortal world, continues to have an existence, and takes an interest in the well-being of subsequent generations (Lantz 2009, Liu 2003). Ancestors are venerated for the foundation they helped establish for the living generations, and are perceived to look out for the interests of their descendants if treated with respect, remembrance and reverence. The spirits of people who died a violent death may be malignant rather than benevolent unless given appropriate rites and offerings, and those who were lost at sea and never given proper funeral rites may wander endlessly as restless spirits unless, perhaps with the intercession of a whale, they are brought to shore.[93] As the whale is seen as an ancestor of fishers, it is not only treated with due reverence and veneration, but it is also perceived to bestow good fortune on the lives and livelihoods of those who believe in the whale-genie and propitiate its spirit. As such, there is no rational tension between spiritual and utilitarian aspects.[94]

[91] While the discussion in this section focuses principally on so-called 'traditional societies', the veneration of whales is quite clearly not restricted to this particular context. According to Kalland (1994), the whale has become – although some Vietnamese, Māori and others consider it always to have been – 'our counter-part in water' and 'whales are made an object of cult-like worship of groups caught in the New Age movement'; one of the utilitarian uses for whales by humans now is in whale-watching, which is worth many millions of dollars to the tourism industry.

[92] I received the same reaction to my apparently naïve attempt to ascertain whether people venerated the whale-genie out of respect or fear – in the sense that failure to make appropriate offerings and participate in appropriate ceremonials would increase the risk of danger while at sea. The elders stated quite emphatically that people venerated whales out of love and respect, not out of fear.

[93] In Da Nang there are two temples to the spirits of fishers or seafarers who have been lost at sea.

[94] Taylor (2004) suggests that modernisation and cultural erosion in Việt Nam are reducing people's genuine spiritual connection with deities, and instead increasing their pursuit of personal gain from the offerings they make to spirits and gods.

There are many references in the literature to 'rescue narratives' – stories of fishers being saved from death by the benevolent actions of whales (see for instance Huy *et al.* 2003, Lantz 2009, Small 2009, Smith *et al.* 1995). I was told a story of a man who was lost at sea for four days in a tiny coracle, without drinking water and shelter (normally coracle fishers will stay in inshore waters and only be at sea for a matter of hours, usually overnight if fishing for squid). All in his village were certain he had died, but he said he was delivered to the shore on the back of a whale, neither hungry nor thirsty. Another man in his 70s, from Nam O village, was shipwrecked some seven kilometres from shore during a typhoon, his boat sunk to the bottom of the sea, but he was delivered unconscious to the shore by a whale. He said that he only had a feeling of warmth. A village elder told me, in response to my question about the utilitarian focus of whale worship:

> ...this case is not the reason why we worship the whale. It just reconfirms our belief. Whales helping people is genuine. This phenomenon happened very long ago, everywhere, and also recently; we witnessed a case by ourselves. The man came to the coast and he was lying on the body of a whale. So that means that the whale supports, rescues the fishing people.[95]

One of the more remarkable rescue narratives which was related to me at several whale temples in the Da Nang-Hội An corridor concerned the escape of so-called 'boat people' from the area ahead of the advance of communist forces from the north of Việt Nam during the period of reunification around 1975. People escaped in whatever craft they could gain access to, including tiny coracles, and many of the established seafarers from the area prayed to Ca Ông for safe deliverance. Those who survived the perils of the oceans, and who in most cases ended up in distant lands – making up communities of Việt Kiều or overseas Vietnamese – believe that the whale-genie was responsible for their safe passage, and they continue to venerate the whale. Most of the whale temples that I visited have tablets on the walls listing their more important benefactors, and several of the more sizeable donations have come from former community members now living in the USA, Australia and Canada. Financial contributions are received from abroad each year for the memorial day festivities, and more and more Việt Kiều are returning to participate physically in the ceremonials.

One of the most famous rescue narratives in Việt Nam – although it is neither possible to establish the extent to which it may be a myth, nor the degree to which it has influenced the narratives that are still heard today – concerns Prince Nguyễn Phúc Ánh, who was later to become Gia Long, the first Emperor of the Nguyễn

[95] Village elder, Nam O village, personal communication, 6 April 2011.

Dynasty that ruled Việt Nam from 1802 until 1943. Whilst the Nguyễn family had long been influential in southern Việt Nam from the mid-16th to the late-18th century, the Nguyễn Dynasty was only established after putting down the peasant-based Tây Sơn Uprising which lasted from 1777 to 1802. During a battle in 1783, the forces of Prince Nguyễn Ánh were forced to flee by sea and, near Phú Quốc Island, were hit by a violent storm. As the prince's boat started sinking desperate prayers were offered to the spirits of the sea, where after a whale is reputed to have positioned itself under the boat, calming the seas surrounding it and carrying it and the men it contained safely to the shore (Lantz 2009, Ruddle & Tuong 2009, Whale temple keeper, Sơn Trà, Da Nang, personal communication, 4 April 2011). In gratitude, the future Emperor decreed that thereafter all fishing communities must venerate the whale and construct temples dedicated to its worship, and the whale itself was given the title *Cá Ông Nam Hải cú tộc ngọc long Chi thần* [Lord Fish, god of the old jade dragon of the South Seas] and, as was common practice at the time, given high-level official designations, such as High Grade Mandarin (Ruddle & Tuong 2009). In 1858 the whale was given the title Admiral of the South Seas, and was called variously by Nguyễn Emperors *Nhân Ngu* [altruistic fish] and *Đức Ngu* [virtuous fish]. The historical *Monograph of the Great Unified South* [*Đại Nam nhất thống chí*], compiled between 1865 and 1875, and completed in 1882, refers to the whale which '…regularly comes to the rescue of shipwrecked people' (Kiem 1996: 147).

This historical episode raises questions as to whether or not whale veneration was 'invented' at this moment, which would put a slightly different gloss on whale worship as a traditional form of ancestor worship among fishing communities in central-south Việt Nam. However, some research (e.g. Durand & Tan 1953, Mus 1975, Thanh 2007) has suggested that the whale-genie may be an evolved form of the deities of the seas which the seafaring Cham people – who inhabited the region of Chăm Pa in what is now southern Việt Nam before its conquest in 1471 by the Đại Việt lord Lê Thánh Tông – venerated long before the establishment of the Nguyễn Dynasty. It is difficult to be categorical because of the fluidity and syncretism with which religious practices functioned, manifested and evolved, but Mus (1975, see also Mus 1931) has suggested that localised Cham ancestor cults existed before Chăm Pa was Hinduised from the 9th century, and that these cults re-emerged after the disintegration of the kingdom from the late 15th century.

Furthermore, elements of the pre-existing Cham culture were readily absorbed by the Việt people who migrated to the region following the Đại Việt conquest. The queen and goddess of the Cham people, Pō Nagar (named Thiên Y A Na by the Vietnamese), is said by legend to have flung herself from a precipice at the time of the Đại Việt conquest of Chăm Pa, only to be rescued by a Hainanese prince (Taylor

2004). But rather than marry the prince out of gratitude rather than love, she threw herself into the ocean and perished. Her body drifted back to Chăm Pa on a log of sandalwood, her body as pure as if she were still alive, and the Cham people erected a shrine dedicated to her veneration. Thereafter, sailors and fishers reported that the goddess appeared to them during violent storms, calming the seas and saving them from death. According to Taylor (2004: 65): 'After taking over the kingdom of Champa, the Vietnamese came to share the Chams belief in her as a protector deity'. In their ongoing migration, they brought the worship of Thiên Y A Na southward. This combination of an historical seafaring Cham people in the central and southern parts of present-day Việt Nam, following ancestor cults, relating legends concerning the rescue of sailors in danger, and the absorption of local practices by the formerly land-based Việt offers a quite plausible historical background to the modern cult of whale veneration in the former Cham region of central-south Việt Nam. Indeed, Kiem (1996: 155) is even more explicit on this:

> During the inevitable march toward the South that began in the mid-11th century (1044), the Vietnamese did not hesitate to adopt a certain number of the beliefs from these peoples [Chams] and to incorporate new deities into their own system of beliefs. Among the newly-found deities figured the whale spirit. ...The Chams revered the whale as the "King of the Waves", and that it was through contact with these people that the Vietnamese became interested in these enormous pachyderms.

The theme of a female protector deity for fishers and seafarers also resonates cross-regionally. The similarities between the stories above and the goddess Mazu (to the Chinese and Taiwanese, or Tian Hou to the overseas Chinese, Thiên Hậu to the Vietnamese) are remarkable. Ma Tsu (Lin Mo) was a young girl, born in 960 on Mei-chou [*Meizhou*] island, Fujian province in mainland China, during the Song Dynasty (960-1279). A devout Buddhist, she died young and unmarried trying to save her father and brothers from drowning. During her lifetime she was believed to have performed many miracles (Ruitenbeek 1999), and after death continued her efforts to save people in danger at sea (Sangren 1988, 1993). Her role in protecting fishers and sailors saw her reputation spread from a localised cult to encompassing the entire southern China seaboard, reaching Guangdong in 1240 and Tianjin in 1326, before then spreading to Ryūkyū in Japan, and then on to South East Asia by the early 18th century – significantly, some time *before* the rescue narrative of

Prince Nguyễn Ánh.[96] She not only helped to save ordinary seafarers, but was also attributed with having protected high-level officials, such as the great explorer Zheng He who led several maritime expeditions between 1407 and 1433 (Lombard-Salmon 1969), and this led to Mazu's official recognition as *tianfei* [Heavenly Princess], *shengmu* [Holy Mother] and then in 1683 *tianhou* [Heavenly Empress], in part for helping the Qing conquest of Taiwan (Antony 2003). Mazu is now the second most important deity in China. Sailors would take onto their boats incense embers from their Tian Hou temples and suspend them in front of the shrines to the deities they worshiped (Antony 2003). They also harboured certain beliefs and superstitions about the creatures of the sea:

> Sailors believed that certain sea creatures, such as sturgeon, sawfish, whales, and sea turtles, were sacred and should not be caught. To avoid calamity, if a sacred fish was caught and killed it was taken to a temple and offered as a sacrifice... Fishermen... took pains not to offend sacred sea creatures and held elaborate ceremonies and sacrifices to assure themselves safety at sea and abundant catches (Antony 2003: 151-152).

One of the more remarkable stories of the interaction between aquatic mammals and humans concerns cooperative fishing with dolphins in Cambodia and elsewhere in the Mekong Basin. Smith *et al.* (1997) report on fishers' description of this very unusual fishing method: Irrawaddy dolphins could be called to their boats by various banging or vocal sounds; the dolphins would slap the water with their flukes if they were willing to cooperate; they would herd shoals of fish by swimming in ever decreasing circles (as they do when seeking fish for themselves); the dolphins would then indicate when the fishermen should cast their nets. Somewhat sceptical, Smith and his team were later able to observe this cooperative fishing method directly. They also observed something similar in Myanmar, and heard reports of Irrawaddy dolphins guiding fish into the nets of Thai fishers (Stacey & Leatherhead 1997). There are also reports from the marine environment in Cambodia of beach seiners who follow dolphins and take advantage of their natural collective fishing methods (Smith *et al.* 1997). Tales similar to those of whales being the 'friends of people' are also told for dolphins. In Cambodia dolphins are believed to help people at sea if they are in trouble (Baird & Mounsouphon 1994), warn of approaching storms by their behaviour (Beasley & Davidson 2007), and 'placing the bones of marine mammals in front of the house will protect inhabitants from bad spirits' (Beasley & Davidson

[96] A 'rescue narrative' attributed to Mazu which has very close parallels with that from Prince Nguyễn Ánh in Việt Nam, discussed earlier, is related by Ruitenbeek (1999): an official, Lu Yundi, was ordered to go on a mission to Korea in 1122. He encountered a hurricane [typhoon] while at sea, and seven of the eight ships in his convoy were wrecked. Lu's ship survived, and as he prayed to the heavens for protection a goddess dressed in red – which was Mazu's style – appeared above his mast. Lu begged her for help, and the seas and wind adjacent to the boat calmed down amidst the wider tumult. Lu was saved.

2007: 373). According to Beasley *et al.* (2009) reverence for Irrawaddy dolphins in the folklore of the Mekong region stems from belief in the creatures' human origins.

Vietnamese fishers who practice *Lễ Cúng Cá Ông* [whale worship] also observe another 'festival of the fish' [*Lễ hội cầu Ngư*] where, at the beginning of the fishing season they ask the spirits of the sea for a bountiful harvest. In this festival the whale may be placed as a benevolent mediator between fishers and the sea's resources. Such a ceremonial is symptomatic of what Barbara Watson Andaya describes (2006) as an appeal to the spirits to help overcome the uncertainties and dangers of a sea-based life and livelihood. Japanese traditional whalers prayed to Ebisu-gami to deliver successful fishing expeditions and good catches (Naumann 1974). The Ainu, one of the aboriginal groups of Japan, depicted one of their gods of the sea as an Orca [killer whale], and whales were believed to symbolise security and prosperity (Creason 2004). The Makah believed that the mythical Thunderbird was responsible for plucking whales from the ocean and delivering them to their shore (Miller 2000). In the Torres Strait, a successful hunt at sea was taken as an indicator of a harmonious relationship between humans and their protector spirits, even if these spirits were embodied in the prey they had killed (McNiven 2010). In Ghana and Togo, the manatee is seen as the embodiment of a water goddess, and is a potential source of both wealth and beauty (Dodman *et al.* 2008), or in Congo as a mermaid who may grant fishers bountiful catches.

Conclusion

In Việt Nam, the veneration of whales [*Lễ Cúng Cá Ông*] and beseeching a good harvest [*Lễ hội cầu Ngư*] are, in my view, quite different forms of interaction between people and the marine environment, but in some coastal areas they – and thus their spiritual *vis à vis* utilitarian underpinnings – are becoming increasingly difficult to separate, particularly where the tourism industry has latched on to these traditional and often quite colourful practices as a spectacle for both foreign and domestic tourists to enjoy.[97] As Lantz concludes in her book on whale worship in Việt Nam, and as I have recently explored further (Parnwell 2013), '…the greatest enemy of Cá Ông worship is, sadly enough, tourism' (Lantz 2009: 55). The intimate and intricate

[97] For example, in September 2009 the Deputy Director of the Ho Chi Minh City Department of Culture, Sports and Tourism indicated the district's intention to 'turn the *Nghing Ong* [whale worship] Festival into a major tourist event'. Available at: http://en.vietnamplus.vn/Home/Fishermen-prepare-for-four-days-of-whale-worship/20099/2900.vnplus. Similarly, an article published on 23 May 2011 in the VietNamNet online newspaper, entitled 'Too Many Sea Festivals in Vietnam', complained that the combination by the tourism authorities of the traditional Cau Ngu [fish prayer] festival with the Cua Lo Tourism Festival in Nghe An Province was turning visitors away because of the emphasis of modern tourism entertainments which detracted from the traditional Cau Ngu activities. Available at: http://english.vietnamnet.vn/en/travel/8384/too-many-sea-festivals-in-vietnam.html.

relationship between people, spirits and the creatures of the sea is facing pressure as development, modernisation and cultural change threaten to transform these traditional cultural practices as coasts are built upon, marine resources overexploited, the sea is polluted, fishers leave for other occupations, and the young generation shows little interest in following the beliefs and practices of their elders, and their ancestors. This is not to advocate the freezing and fossilizing of local cultures, which must evolve to survive, but to emphasize – as I hope this chapter has shown – that there are spiritual sensitivities, structures and sinews that penetrate deeply into societies but also extend widely across space which remind us of our dependence on and responsibilities towards nature, the ostensibly beneficent creatures with whom we cohabit this planet, and one another.

This chapter has also sought to draw attention to 'regional resonances' in people's spiritual and historical connections with the sea, and with sea mammals and creatures as distant ancestors, which reverberate across the Asia-Pacific and beyond – broad similarities in the manifestations of and motivations behind the veneration of sea spirits and deities. The sea is a dangerous place for fishers and sailors, and its resources are not easily obtained. The propitiation of deities and spirits reflects traditional societies' respect for (and occasionally fear of) the intangible forces which mediate their relationship with the sea. Deep within these spiritual practices might be found a conservation ethic wherein the limits of resource exploitation might be defined and an appreciation of the marine ecosystem operationalised. It is disconcerting that mounting efforts are taking place at the local, national and regional levels which are intensifying development pressures on and the transformation of marine and coastal environments, whereas only minimal recognition is given to the ways that local communities and their traditional beliefs and practices have historically underpinned a far more harmonious relationship between people, the sea and its resources. Since 2003 UNESCO has recognised the potential contribution of 'intangible cultural heritage' to sustainable development.[98] Moreover, a growing number of states parties are becoming signatories to the Convention for the Safeguarding of Intangible Cultural Heritage. Hitherto, only one manifestation of intangible cultural heritage which relates to coastal communities' and seafarers' relationship with the sea has been recognised in the Asia-Pacific region: Mazu belief and customs in China were inscribed on the Representative List in 2009 as an historic manifestation of coastal communities' identification with this goddess of the sea. There are some 5,000 Mazu temples across the world, most of them in East and South East Asia, and yet only in one country, China, are these institutionally recognised and protected, notwithstanding the cultural and spiritual sinews which

[98] Available at: http://www.unesco.org/culture/ich/index.php?lg=en&pg=00006.

reach out and connect seafaring communities cross-regionally. Whilst strenuous efforts have been made to foster regional cooperation in the interests, *inter alia*, of modernisation, development and security, the soft cultural underbelly of regionalism has been shamefully overlooked. There is an urgent need for joined-up thinking and bottom-up action with respect to the nexus of development, culture and sustainability in the Asia-Pacific region, for instance in the safeguarding of intangible cultural heritage across and beyond national boundaries. Regionalism is not simply an issue of international political economy; it is a lived reality in a region where the sea has historically been the principal mode of cultural interconnectivity.

Bibliography

Acebes, J.M.V. 2009, Historical Whaling in the Philippines: Origins of 'Indigenous Subsistence Whaling', Mapping Whaling Grounds and Comparison with Current Known Distribution: A HMAP Asia Project paper', *Working Paper No. 161, October 2009*. Asia Research Centre, Murdoch University, Perth, Australia.

Andaya, B.W. 2006, Oceans Unbounded: Traversing Asia Across 'Area Studies ', *The Journal of Asian Studies*, 65(4): 669-690.

Antony, R.J. 2003, *Like Froth Floating in the Sea: The World of Pirates and Seafarers in Late Imperial South China*, Institute of East Asian Studies, China Research Monograph 56, University of California, Berkeley, CA, USA.

Baird, I.G. & Mounsouphom, B. 1994, Irrawaddy Dolphins (*Orcaella Brevirostris*) in Southern Lao PDR and Northeastern Cambodia, *Natural History Bulletin of the Siam Society*, 42: 159-172.

Barnes, R.H. 1974, Lamalerap: A Whaling Village in Eastern Indonesia, *Indonesia*, 17: 136-159.

Barnes, R.H. 1996, *Sea Hunters of Indonesia: Fishers and Weavers of Lamalera*, Clarendon Press, Oxford, UK.

Barnes, R.H. 2005, Indigenous Use and Management of Whales and Other Marine Resources in East Flores and Lembata, Indonesia, *Seri Ethnological Studies*, 67: 77-85.

Beasley, I.L. & Davidson, P.J.A. 2007, Conservation Status of Marine Mammals in Cambodian Waters, Including Seven New Cetacean Records of Occurrence, *Aquatic Mammals*, 33(3): 368-379.

Beasley, I., Marsh, H., Jefferson, T.A. & Arnold, P. 2009, Conserving Dolphins in the Mekong River: The Complex Challenge of Competing Interests. In: Campbell, I.C. (ed.). *The Mekong: Biophysical Environment in an International River Basin. Aquatic Ecology Series.* Elsevier, New York, NY, USA: 363-387.

Creason, A.M. 2004, Culture Clash: The Influence of Indigenous Cultures on the International Whaling Regime, *California Western International Law Journal*, 35(1): 83-126.

David, B., McNiven, I.J., Crouch, J., Skelly, R., Barker, B., Courtney, K. & Hewitt, G. 2009, Koey Ngurtai: The Emergence of a Ritual Domain in Western Torres Strait, Archaeology in Oceania, 44: 1-17.

Dodman, T., Diop, N.M.D. & Khady, S. (eds.) 2008, *Conservation Strategy for the West African Manatee*, UNEP, Nairobi, Kenya and Wetlands International Africa, Dakar, Senegal.

Durand, M. & Tan, T.H. 1953, Culte de la Baleine: Chant des pêcheurs de Truong Dong [The Cult of the Whale: Songs of the Fishermen of Truong Dong], *Bulletin of the Society of Indochinese Studies*, XXVIII(2): 183-219.

Huffington Post 2010, 'Your Excellency' Whale Mourned: 10,000 Give Last Rites to 15-Ton Dead Whale in Vietnam, *Huffington Post*, 23 February 2010. Available at: http://www.huffingtonpost.com/2010/02/23/your-excellency-whale-mou_n_473155.html.

Huy, N.V., Ngoc, N.A., Hong, N.H. & Dung, N.T. 2003, The Village God's Journey. In: Huy, N.V. & Kendall, L. (eds.). *Vietnam: Journeys of Body, Mind and Spirit.* University of California Press, Berkeley, CA, USA: 216-237.

Kalland, A. 1994, Super Whale: The Use of Myths and Symbols in Environmentalism. In: The High North Alliance, 11 Essays on Whales and Man, 2nd Ed., 26 September 1994. Lofoten, Norway.

Kiem, T.V. 1996, The Cult of the Whale, *Vietnamese Studies*, Special Issue on Religions and Popular Beliefs in Vietnam, 51 NS (121 OS)(3): 145-172.

Lantis, M. 1938, The Alaskan Whale Cult and its Affinities, *American Anthropologist*, 40(3): 438-464.

Lantz, S. 2009, *Whale Worship in Vietnam*, Religionsvetenskapliga studier från Gävle 4. Swedish Science Press, Uppsala, Sweden.

Liu, T. 2003, A Nameless But Active Religion: An Anthropologist's View of Local Religion in Hong Kong and Macau, *The China Quarterly*, 174: 373-394.

Lombard-Salmon, C. 1969, La Communauté Chinoise de Makasar, Vie Religieuse, *T'oung Pao*, 55(4-5). Brill, Leiden, the Netherlands: 241-297.

Macdonald, C.J.H. 2003, Le culte de la baleine, une exception vietnamienne?, *Aséanie*, 12: 123-136.

McNiven, I. 2004, Saltwater People: Spiritscapes, Maritime Rituals and the Archaeology of Australian Indigenous Seascapes, *World Archaeology*, 35(3): 329-349.

McNiven, I.J. 2010, Navigating the Human-Animal Divide: Marine Mammal Hunters and Rituals of Sensory Allurement, *World Archaeology*, 42(2): 215-230.

Miller, R.J. 2000, Exercising Cultural Self-Determination: The Makah Indian Tribe Goes Whaling, *American Indian Law Review*, 25(2): 165-273.

Morgan, R.C. 2007, Property of Spirits: Hereditary and Global Value of Sea Turtles in Fiji, *Human Organization*, 66(1): 60-68.

Mus, P. 1931, Etudes indiennes et indochinoises: IV: deux légendes chames, *Bulletin de l'École Française d'Extrême-Orient*, 31(1): 39-102.

Mus, P. 1975, India Seen from the East: Indian and Indigenous Cults in Champa, Monash Papers of Southeast Asia, no. 3, Centre of Southeast Asian Studies, Monash University (originally published as 'Cultes indiens et indigènes au Champa ', *Bulletin de l'Ecole Française d'Extrême Orient*, xxxiii, 1993: 367-410).

Naumann, N. 1974, Whale and Fish Cult in Japan: A Basic Feature of Ebisu Worship, *Asian Folklore Studies*, 33(1): 1-15.

Parnwell, M.J.G., Whale Worship and Tourism Development in the Hội An-Da Nang Corridor, Việt Nam, *South East Asia Research* (forthcoming, summer 2013).

Ruddle, K. & Tuong, P.L. (eds.) 2009, The *van chai* of Việt Nam: Managing Nearshore Fisheries and Fishing Communities, *International Resources Management Institute Publication*, no. 3, Hong Kong, China.

Ruitenbeek, K. 1999, Mazu, The Patroness of Sailors, in Chinese Pictorial Ar', *Artibus Asiae*, 58(3/4): 281-329.

Sangren, P.S. 1988, History and the Rhetoric of Legitimacy: The Ma Tsu Cult of Taiwan, *Comparative Studies in Society and History*, 30: 674-697.

Sangren, P.S. 1993, Power and Transcendence in the Ma Tsu Pilgrimages of Taiwan, *American Ethnologist*, 20(3): 564-582.

Segi, S. 2003, The Coexistence of Whaling and Whale Watching in a Traditional Whaling Region: The Case of Taiji, Wakayama Prefecture, Japan, *SPC Traditional Marine Resource Management and Knowledge Information Bulletin*, no. 15: 21-25.

Shindo, N. 1975, *History of Whales in the Inland Sea*. Dietists' Institute, Hyogo, Japan.

Small, I.V. 2009, *Beyond the Sea: Geographies of Mobility and Memory in Quy Nhon, Viet Nam*, proceedings of 'Diasporas: Exploring Critical Issues', 7 June 2009, Mansfield College, Oxford, UK. Available at: http://www.inter-disciplinary.net/at-the-interface/diversity-recognition/diasporas/project-archives/2nd/session-ii-creating-consciousness/.

Smith, B. D., Jefferson, T. A., Ho, D. T., Leatherwood, S., Van Thuoc, C., Andersen, M. & Chiam, E. 1995, Marine Mammals of Vietnam: A Preliminary Checklist, *Collection of Marine Research Works*, VI: 147-176.

Smith, B.D., Jefferson, T.A., Leatherwood, S., Ho, D.T., Van Thuoc, C. & Quang, L.H. 1997, Investigations of Marine Mammals in Vietnam, *Asian Marine Biology*, 14: 145-172.

Smith, B.D., Thant, U.H., Lwin, J.M. & Shaw, C.D. 1997, Investigation of Cetaceans in the Ayeyarwady River and Northern Coastal Waters of Myanmar, *Asian Marine Biology*, 14: 173-194.

Stacey, N., Karam, J., Dwyer, D., Speed, C. & Meekan, M. 2008, Assessing Traditional Ecological Knowledge of Whale Sharks (*Rhincodon typus*) in eastern Indonesia: A Pilot Study with Fishing Communities in Nusa Tenggara Timur, *Report prepared for the Department of the Environment, Water, Heritage and the Arts*, School of Environmental Research, Charles Darwin University, Canberra, Australia.

Stacey, P.J. & Leatherhead, S. 1997, The Irrawaddy Dolphin, *Orcaella Brevirostris*: A Summary of Current Knowledge and Recommendations for Conservation Action, *Asian Marine Biology*, 14: 195-214.

Sullivan, R. 1998, Permission Granted to kill a Whale. Now What?, *New York Times*, 9 August 1998.

Taylor, P. 2004, *Goddess on the Rise: Pilgrimage and Popular Religion in Vietnam*, University of Hawai'i Press, Honolulu, HI, USA.

Thanh, N.T.Q. 2007, Le Culte de la Baleine dans le Centre Viet-Nam devenir d'un héritage multiculturel, *Péninsule*, 55(2): 97-125.

Chapter 6: The march of the monks

On the political repertoire of Buddhist monks in Myanmar and Sri Lanka

Michael Hertzberg

In September 2007 Myanmar witnessed the uprising of thousands of monks who marched the streets of Yangon column by column in protest against the long-standing military regime. At its peak an estimated 100,000 monks marched in the streets of Yangon, and a total of 300,000 monks and nuns marched throughout the country in one of the few public acts of mass protests since the 1988 uprising (Schober 2011). The march of the monks went on for days. It came to dominate the headlines of international media, which impatiently asked: 'Will this be the end of the military regime?' In 2004, a new political party named Jathika Hela Urumaya emerged on the political scene, consisting only of Buddhist monks. Their agenda was to end the Norwegian-led peace negotiations and introduce anti-conversion legislation in Sri Lanka. My aim is to find out whether we could posit a unique political space open for the Buddhist clergy, irrespective of their various ideologies they may pursue, and to see how the respective political repertoires of Buddhist monks unfold in different national contexts.

The prelude to the uprising of the monks in Myanmar was the military regime's proposed constitution in the spring 2007. The proposal would have secured the continuing role of the military in Burmese politics. Several minor and sporadic protests were held in Myanmar, known for its brutal repression of political freedom. After two monks were beaten up in Pakokku, the formation of the All Burma Monks Association (ABMA) emerged on the numerological auspicious date of 9th September 2007 (9 is seen as an enchanted number) demanding an apology from the regime. The leader of ABMA, U Gambira, was to take a prominent role in the upcoming uprising (Gravers 2012, Lintner 2009). However, it was the persistently high levels of inflation that eventually were to provoke an uprising. In 2007 consumer prices rose by over 35%, adding to the burden of a 20% increase the previous year.[99] Yet it was the sharp rise in the price on fuels and natural gas that triggered the monks

[99] Data from the World Bank. Available at: http://search.worldbank.org/data?qterm=inflation%20 myanmar&language=EN. It should also be noted that the World Bank Data are based on official figures. The economy of Myanmar is, however, characterised by its enormous informal market. Thus, these inflation figures are likely to underestimate the increase of the cost of living for the population at large.

C. Brautaset and C.M. Dent (eds.) **THE GREAT DIVERSITY**
DOI 10.3920/978-90-8686-786-8_6, © Wageningen Academic Publishers 2014

to turn to the streets.[100] The monks' intimate relations with the poverty of the lay people in Myanmar, made their march a direct challenge to the rule of the military regime. The regime did not provide an apology for the violence against the monks in Pakokku, and thus ABMA issued a four-point statement demanding the release of political prisoners, price cuts to combat inflation and political reforms. When these demands also were refused, they issued a second statement calling for a boycott of the regime, by launching an alms-boycott campaign [*Patta Nikkujjana Kamma*] (Shen & Chan 2010). In a recent publication, Juliane Schober describes the march 'as an army of young men in robes who had taken to the streets on a pilgrimage of justice' (Schober 2011: 125).

The route of the march was not arbitrary, but followed the tracks of earlier protest marches.[101] Other destinations included sacred sites such as the Shwedagon Pagoda and Sule Pagoda. Thus, these political routes were imbued with a sacred aura, but it also made these sacred sites concomitant with political opposition. The march also took an unexpected turn into University Avenue, where Myanmar's lady of democracy, Aung San Suu Kyi, was held captive. The meeting between the monks and Aung San Suu Kyi was brief, but still of high significance. University Avenue is usually closed by the military, yet the demonstrators had been able to strike a deal with the local commander to attain permission for a meeting. By the regime's tacit permission to march for several days and display other forms of open opposition, people in Myanmar were led to believe that change was about to come. These marches have retrospectively been referred to as the Saffron Revolution.

The military regime, however, began its violent crackdown on the demonstrations 26th of September 2007. The regime was particularly focussed on tracking down leaders from the protests, among them several monks. Estimates of total arrests vary between 3,000 to 6,000, and the figures of fatalities between 31 (government figures) and 200 (human rights groups). The compounds of monks were visited and ransacked at night, and many of these buildings were emptied except for elder monks who were looking after the compounds. Many monks fled, either to their village of origin, or over to the Thai border. U Gambira was first given a sentence for 68 years after being defrocked and re-ordained in the same manner. An unknown number of monks were defrocked as part of the process of harsh interrogations.

[100] The cost of petrol and diesel doubled, and the price on natural gas fivefold within a week. The accelerating inflation was due to the regimes decision to cut subsidies of fuel. It should also be mentioned that Myanmar has been a net exporter of natural gas since 2002, and that gas currently accounts for around 40% of its export earnings.

[101] Most notably from 8/8/88, but also along routes dating back to the anti-colonial protests.

The distinction between the secular and monastic law in Myanmar offer its own unique dynamics. Why did the *tatmadaw*, the military, need to defrock the monks to interrogate them? Legally, the monks in Myanmar are not subject to the civil laws of the country, but rather to the *vinaya*, the monastic laws only. However, if the nature of the monk's transgression is of such nature that he is expelled from the *sangha*, he is again a subject under the civil law. As such, there are two parallel systems of law in Myanmar, and the interference of the military into the sphere of the monastic laws is considered offensive (Schober 2011). Reports that several monks have initiated hunger strikes in detention have emerged, and this resembles the anti-colonial protests of the U Wisara who died in British detention after a prolonged hunger strike (Smith 1965). Thus, the activities of the monks during the Saffron Revolution also play upon a greater historical awareness of political resistance.

Colonial and postcolonial activities of political *pongyis*

Burma came under British rule after the third Anglo-Burmese War in 1885, and one of their main concerns was that a systematic involvement and consolidation of the Buddhist *sangha* would enter into the anti-colonial resistance. However, such national sentiments and anti-colonial resistance would not primarily come from the *sangha* hierarchy, but from various political *pongyis* [monks], who were relatively independent from the hierarchy of the *sangha*. One of the first religio-political organisations to form in Burma was the nationalist *Sangha* Sametggi Association. The British and the *sangha* hierarchy issued an order in 1921 that the organisation should abstain from political agitation (Smith 1965).

According to the monastic law [*vinaya*] there is a sharp distinction between mundane affairs [*laukika*] and other-worldly nirvana-seeking efforts [*lokuttara*]. This divide has often been the main point of exploration into the dynamics of monks and politics. In practice, however, this divide has been less rigid. In her book *Modern Buddhist Conjunctures in Myanmar* (2011), Juliane Schober takes as decisive stance against this mundane/sacred divide and rather explores how 'Buddhist conceptions and practices are intimately tied to conceptions of political power in social, economic, and political realm' (Schober 2011: 11). Thus, there is no need to portray a political *pongyi* as a violator of his own precepts, or as a paradoxical figure because of his political involvement. Yet the distinction between *lokuttara* and *laukika* does have a normative element to it, which often infers the discourses and debates around the legitimacy of the political *pongyi*. Aung San, the father of Aung San Suu Kyi and the founder of the Burmese military, had a strong stance against the fusion of religion and politics, and was a vocal opponent of religious involvement into politics

(Gravers 1993). Lintner (2009) reports a variety of different self-designations by monks concerning the political nature of the 2007 uprising.

One element of this is precisely the designation 'political *pongyi*'. Schober claims that terming monks as political '…made scholars complicit in the hegemonic discourse of the state' (Schober 2011: 139). The designation is also used as a way of discrediting the engagement of monks in political matters as a transgression of their *vinaya* precepts. British colonial authorities introduced the designation in the 1920s in an attempt of diminishing the legitimacy of actions taken by activist monks such U Ottama and U Wisara and other anti-colonial Buddhist formations. Schober (2011) argues that the military regime follows the same means in designating such monks as 'political monks' to deprive them of their religious sincerity. However, my own usage of the concept stem from Carl Schmitt's assertion that '…all political concepts, images, and terms have a polemical meaning' (Schmitt 2007: 30). Thus, I am more interested in the nuances and remembrances such a concept may evoke (also in the Sri Lankan context) rather than construing an analytical meta-language in talking about religion and politics among Buddhist monks as such. As we will see in the Sri Lankan context, it is not only the Burmese state, but also a frequent claim from the *sangha* hierarchy that monks should not engage in political work. Most often the designation 'political monk' will be derogatory, yet the precise task of political monks is to negotiate this claim either in terms of wording (as activist monks) or as a state of exception (due to a crisis). By using the term 'political monk' [or *pongyi*], this threshold between legitimacy and illegitimacy needs to be explained, and cannot be sidestepped.

However, the first massive Buddhist protest against the British came with the so-called 'no footwear-issue'. This concerned the practice of Europeans keeping their shoes on while visiting pagoda premises, contrary to the local customs of removing ones shoes. In 1916 The Young Men's Buddhist Association brought forward the issues with the government, yet their request was met with a ban on further public discussion of the matter. It was not before 1919 that the British conceded on the issue of no-footwear at pagodas, but the issue had already evoked Buddhist-nationalist sentiments among the Burmese, in which political monks were to take a prominent role (Smith 1965).

U Ottama came back from India in 1921 and brought with him several ideas from the Indian national resistance and the non-cooperation movement. His motto could be subsumed into: first freedom, then nirvana. He was sent to prison for the first time in 1921 after delivering a set of fiery speeches attacking the British colonial rule. From then onwards, U Ottama was to endure several prison sentences until his death in 1939. His arrest led to an upsurge of popular resistance against the colonial rule, and massive actions of non-cooperation were initiated. The next decade saw an increasing participation of monks in political affairs. Another monk from this period that is still publicly venerated is U Wisara, who died in prison in 1929 after commending a prolonged hunger strike in protest of the refusal to let him wear his yellow robe whilst in prison. Similar protests have been undertaken by monks that have been held as political prisoners by the regime, directly inspired by the anti-colonial efforts of U Wisara.

Donald Eugene Smith argued already in 1965 that '...the monks were thus in a uniquely advantageous position to participate actively in nationalist politics' (Smith 1965: 64). Several characteristics of a unique political space for the Buddhist clergy can be identified both in historical and contemporary settings. The robe entitled them to both social prestige and an enormously influential position within society, of which they could draw upon for political agitation. Secondly, the monks were organised into temples, which were further organised into *kyaungtaiks* (cluster of monasteries). Hence, they could mobilize for demonstrations, public meetings and processions quickly and their networks reached into the tiniest village in utmost rim, a feat no political party could match (Smith 1965). Thus, the *sangha* can be seen as a parallel structure to the military, due to their organizational capacity both in local, regional and national outreach. These networks have enabled the *sangha* to be a continuous source of opposition to the regime, as well as ensuring that the potential to engage effectively in an uprising is ever present. Other forums of oppositional activities are the literary clubs, education classes and teahouses (Larkin 2011, Lintner 2009). As the regime does not allow any substantial collective gatherings, especially those of political character, different forms of networks play a significant role in maintaining the opposition to the regime. Several monks and other political opponents to the regime have fled to the nearby countries (Thailand, India and Sri Lanka) where they continue their political work (Lintner 2009, Shen & Chan 2010).

The monks could also enjoy relative freedom from the various bodies of law, both ecclesiastical and secular. Whilst monks in Thailand are directly prohibited by the *Mahasangha*, the supreme ecclesiastical authority, to engage in political affairs, no such supreme entity exists in Burma/Myanmar. Regarding secular law, the authorities are reluctant to prosecute monks, especially due to the widespread indignation such actions invariably provokes, which is equally true with the British administration (Smith 1965) as with the present military regime (Schober 2011). However, this respect for the monks has not prevented the regime from conducting violent crackdowns on the clergy.

Another feature of the political potential of the *pongyi* is their lack of vulnerable relations that would be jeopardised by potential political involvement; there were no jobs, families or possessions that could be taken from them (Smith 1965). The same argument figures in the present situation: 'We monks have nothing to lose' Schober quotes a monk saying in relation to the responsibility of the monks vest into the resistance against the regime on behalf of the common people (Schober 2011: 136). However, political *pongyis* has been banned from direct political participation in elections since 1936, and they are not allowed to register their vote either. As we will see later, this is in direct contrast with the situation in Sri Lanka, where monks are even allowed to run for office in the political sphere.

While monks in Burma/Myanmar have been barred from entering politics directly, there are numerous examples of their interaction in politics through various pressure groups. The usual procedure of these groups is to hold a conference or seminar, and then issue a resolution which is forwarded to the government. However, more desperate measures have also been applied. From 1955 onwards, there was an on-going discussion whether Buddhism should be declared the state religion of Burma. The first Premier of Burma, U Nu, who has been widely accused for favouring Buddhism and holding close ties with the religious elite, opened for making Buddhism the state religion. However, when he in 1961 tried to enact an amendment to this declaration that would strengthen the minorities' rights to religious freedom, more than 2,000 monks surrounded the premises of the government and coerced the legislators from voting for the amendment (Smith 1965: 196). When Ne Win instigated the military coup in 1962, he quickly severed the tight bonds the government of U Nu had established with Buddhism.

One aspect that is often highlighted with regards to the situation in Burma/ Myanmar is that while it is possible to ignore the *sangha* in political matters, it is unwise to displease them. As such, even the military regime has since the 1980s tried to establish closer ties with the *sangha*. Still, this did not prevent monks from being at the forefront in the demonstrations in 1988 and in 1990 (Matthews 1993). The military regime has the last decades evolved its staunch division of religion and politics to being more accommodative to religious affairs, and thus become a participant in what Ingrid Jordt terms as 'the *dāna* economy' (Jordt 2007: 97). Dāna, which is the offering made towards monks (e.g. food), is also the main method for accruing merit for a Buddhist layman. The participation of the military regime into the dāna economy have been seen as a conspiratorial move to entail Buddhist legitimacy of their rule. However, Ingrid Jordt argues that this practice rather can be seen as an element in a politics of sincerity, where the monks and the laity are the final judges of that sincerity. For example, she relates to a story where her informants express disappointment when the regime has been able to refurbish the famous Shwedagon Pagoda, an accomplishment that may only be undertaken by a legitimate ruler. Nevertheless, Jordt warns against the perception that the military regime can be dealt with as a monolithic entity alone, and argues that the intentions of individual people also come into play. The *dāna* economy leads us into one of the most powerful political weapons that is employed by the *sangha*, namely the *pattam nikkujjana kamma*, or the turning of the bowl. By refusing donations from unworthy persons, the monks prevent them for accruing merit, and in a way excommunicating them from the *dāna* economy (Jordt 2007).

The action of turning one's bowl is a powerful protest of the monks, an utmost form of political resistance. Kovan argues that now a whole new generation has been politicized (Kovan 2012). When four monks refused to accept alms from military personnel after the landslide democratic victory of Aung San Suu Kyi in 1990, they were shot dead (Jordt 2007). Schober observes that the 'enactment of the *pattam nikkujjana kamma* may therefore harbour a great deal of social tension and potential communal violence' (Schober 2011: 124), and lists that other such collective protests of turning the bowl have taken place in Myanmar in 1990, 1996, 1997, 2004 and in the Saffron Revolution in 2007. It was after a violent episode towards two monks in early September 2007 that the monks decided to field a nationwide campaign of turning the bowl towards the regime and their subsequent families. Along their march in the street, the monks held their bowls upside down to visualize the political boycott of the regime. In stark contrast to this, they proceeded to accept alms from their various lay supporters along the route. The regime does not take lightly upon such boycotts, and do their utmost to have them lifted (Matthews 1993). However, Kovan (2012: 124) argues that: 'presumably any multi-repeated intervention of the *pattanikujjana* risks diminishing its force as a form of ethical leverage merely through over-use; thus, its timing but also frequency are crucial for its efficacy'.

Sri Lanka: the heritage of the political *bhikkhu*

Shortly before the election in Sri Lanka in April 2004, numerous Buddhist monks decided to field their own party, Jathika Hela Urumaya (JHU), to participate in the upcoming election. In total 286 Buddhist monks fielded as candidates for the general election, which is the first time a party exclusively consisting of Buddhist monks decided to contest an election (Deegalle 2006). One of the first things the newly formed party did was to stage a *pāramitā perahera* [virtuous procession] from the Kelaniya temple to the Sri Dalada Maligawa [The temple of the Tooth] in Kandy (Frydenlund 2011). These temples rank high in the consciousness of the Sri Lankan Buddhists, and the potency of the Sri Dalada Maligawa is attested by the fact that the inaugural ceremony of presidency is held there. Thus, the march of the monks from Kelaniya to Kandy was vested with religious importance, and at the end of the march JHU's election manifesto [*amādahara*] was offered at the Sri Dalada Maligawa (Frydenlund 2011). The fusion of religious and political elements were further stated by the fact that the march was held on the 2nd of March, commemorating the 189th anniversary of the Kandyan Convention, the last kingdom in Sri Lanka to succumb under British rule. Six weeks later JHU was able to secure nine seats in the general election, and emerged on the political scene as Buddhist-nationalist surprise (Deegalle 2006).

One note that arises is the simple fact that Sri Lankan monks, in contrast to Myanmar monks, are able to run for elections. Some individual monks have fielded their candidature in the past in Sri Lanka, but never before in a coordinated effort as a group. However, the debate whether the monks should be allowed to get involved in social and political affairs has been long ongoing. After a statement by the first premier, D.S. Senanayake, in Sri Lanka in January 1946 that monks should not interfere in political matters, a coven of monks gathered and issued *The Declaration of the Vidyalankara Pirivena* (see Rahula 2006: Appendix II). This declaration is a concise statement that allows monks to engage in politics, or activities concerning the welfare of the people, as long as it does not hinder the religious life of a *bhikkhu* [monk]. The statement provoked hostility both from politicians and the press alike, and terms such as *dussela* [unvirtuous] were used to designate political monks, and national newspapers claimed that this was a threat from 'a small coterie of political bhikkhus' and that the government should eradicate their influence (Warnapala 1978: 75). The monks responsible for the statement argued that the vocation of the monks no longer was solely for the attainment of nirvana, but also for the good of the country, nation and religion [*rata, jatiya, agama*] (see Seneviratne 1999).

Walpola Rahula, a prominent scholar monk, who was among the small coterie of political monks, published a pamphlet *Bhiksuvage Urumaya* in 1946 (later translated to *The Heritage of the Bhikkhu*) where he argued that Buddhist monks always had stepped forward whenever the country had been in dire need, a book that was written in reply to the debate on the Vidyalankara declaration. It was not until ten years later, in 1956, that the monks were to realise the potential of engaging in political affairs. In 1951 All Ceylon Buddhist Congress requested the ruling party United National Party (UNP) to inquire into the present status of Buddhism in Sri Lanka, and especially the impact of colonialism upon the religion (Phadnis 1976). The UNP reluctantly agreed to appoint a commission and the report *The Betrayal of Buddhism* was released in 1956 and concluded that Buddhism had been neglected in the colonial period and cut off from the polity (Tambiah 1992). The conclusions of the report captured the imagination of the monks, and the report was one of the main reasons for the emergence of the Buddhist pressure group Eksath Bhikkhu Peramuna (EBP) in February 1956 (Phadnis 1976).

The ruling party UNP was reluctant and evasive on the demands of the EBP, which included the controversial Sinhala-Only language proposal, which hindered Tamils in entering the educational sector in their own language. The UNP itself became a target of antagonism, and the political monks in EBP found a new ally in the newly formed Sri Lanka Freedom Party (SLFP) led by S.W.R.D. Bandaranaike, who was willing to meet their demands and mobilized to introduce the Sinhala-Only language act. Approximately 3,000 to 12,000 monks were involved in the election campaign for SLFP and S.W.R.D. Bandaranaike, an alliance that won a landslide victory in the 1956 elections (Imtiyaz 2010).

A year later, in 1957, S.W.R.D. Bandaranaike sought to remedy the injustices done to the minority Tamils through the language act of Sinhala-Only. He suggested a four-point proposal to introduce legislative provisions for the reasonable use of Tamil language, popularly known as the Bandaranaike-Chelvanayakam Pact. However, this proposal met with ardent opposition, especially from the pressure groups of EBP and Sinhala Jatika Sangamaya (SJS), which also advocated for a more firm Buddhist-Sinhala nexus in Sri Lanka. In April 1958, a sit-in of about 500 people (250 monks) was arranged in front of S.W.R.D Bandaranaike's residence, demanding the abrogation of the pact (Phadnis 1976). Their demands were successful and the pact was abrogated. This period also witnessed the first communal tensions between Tamils and the Sinhalese, and the enactment of the language act of Sinhala-Only was a very potent issue in that regard. Phadnis argues that 'both the EBP and SJS were loosely structured, ad hoc in nature and highly personalized in character' (Phadnis 1976: 273). The emergence of the political monks came to an abrupt halt with the assassination of S.W.R.D. Bandaranaike in 1959 committed by two Buddhist monks, an event that shocked the country.

It was not before 1986 that monks re-emerged as a collective force in political matters. The umbrella organisation Mavbime Surakime Vyaparaya, or 'Campaign for the Protection of the Motherland' worked for the protection of Sinhala-buddhist culture and particularly against attempts of devolution of the country, most natably through the up-coming peace treaty called the Indo-Lanka Accord in 1987. As a coalition of pressure groups, politicians, Buddhist lay organisations and with the endorsement of the *Mahanayakes* [Buddhist *Sangha* leaders], they presented themselves as the voice of the Buddhist opinion (Amunugama 1991: 120). Another group that became affiliated with political monks was the revolutionary Janata Vimukti Peramuna (JVP), who waged a violent insurrection against the government in the late 1980s in protest against the signing of the Indo-Lanka Accord, which was a treaty on the devolution of the country. JVP took the definition of the political monk to a new intensity, by placing the political role of the monks before the religious duties. The political bhikkhus in JVP were seen as a non-formal grouping outside the hierarchical *sangha*, and the new image of the political monk was one who sacrificed his religious duties and his life for the protection of the motherland. The JVP was violently defeated by paramilitary groups led by the government, and monks who were found guilty in JVP activities had to undertake rehabilitation to regain their former monastic identity as a monk, focusing on traditional vocations such as meditation, preaching and chanting (Abeysekara 2002).

Since the Vidyalankara Declaration in 1946 the debate on 'political monks' in Sri Lanka has tended to ebb and flow, depending on the various engagement by particular monks. The decision of the JHU monks to run for election was heavily criticised by different segments of the *sangha*, and one *Mahanayake* stated that they would bring about the destruction of the *Buddha Sasana*, and a popular nickname in Sri Lanka is *Jathika Hela Karumaya* [National Sinhalese Curse] (DeVotta & Stone 2008). De Silva terms the period 1960-1982 as the 'Triumph of Orthodoxy' as this period saw few active political monks, as the government allied themselves with orthodox elements within the *sangha* hierarchy to curb the influence of political monks (De Silva 1993). But this was also a period in which several of the demands of the Buddhist clergy actually were met by politicians. The most notable example of this was the constitutional statement of 1972 that declared that Buddhism was the foremost religion of Sri Lanka. However, with the outbreak of the civil war between the government of Sri Lanka and the Liberation Tigers of Tamil Eelam (LTTE) in 1983, Buddhist monks engaged themselves more and more frequently in political affairs, especially in matters concerning proposals of devolution in the country. In 1997 more than 3,000 monks participated in a mass cursing, which entailed the smashing of coconuts under a Bo tree while reciting curses of a cabinet minister after he had dismissed a report on the effects of Tamil militancy on Buddhism in Sri Lanka (Bastin 2010).

In 2003, following the abrupt death of the famous monk Ven. Gangodavila Soma, who many Buddhist-nationalists claims that was murdered by Christian fundamentalists, two other monks, Ven. Omalpe Sobhita Thero and Ven. Rajawatte Vappe Thera, staged a fast-unto-death campaign. The aim of this campaign was to have the government to formulate legislation to ban 'unethical conversions', which allegedly were committed by Christian organizations in Sri Lanka (Owens 2006). The government agreed to establish a special committee in January 2004, yet before this committee was able to produce any result, the emergence of the Buddhist-nationalist party Jathika Hela Urumaya came about. Later, Ven. Omalpe Sobhita Thero was voted into parliament and launched his own proposal for legislation as a Private Members Bill in May 2004 (Owens 2006). However, the practice of staging fast-unto-death campaigns did not end with the parliamentary entry. In June 2005 Ven. Omalpe Sobhita Thero staged another campaign in protest against the P-TOMS, a joint mechanism between the government of Sri Lanka and the LTTE for the distribution of foreign aid after the tsunami. The practice of activist methods did not end for the monk with the entry into parliament, and remained part of their political repertoire vis-à-vis their political participation in parliamentary politics. The continuance of activist methods entered into a wider debate within the JHU whether to commit themselves to parliamentary politics, or to continue their struggle on an activist basis only (Juergensmeyer 2008).

On the political repertoire of Buddhist monks in Myanmar and Sri Lanka

My quest into the genealogies of political monks in Myanmar and Sri Lanka have been to find out whether we could posit a unique political space open for the Buddhist clergy, irrespective of their various ideologies they may pursue. The focus has been on *how* rather than *why* political monks have engaged in the political sphere (not limited to party politics). How would the presence of their robe hinder or enhance various political activities? The first contrast between the countries is found in the elections, where the Sri Lankan counterpart is allowed both to vote and to run for elections. But that does not mean that Myanmar monks are barred from political engagement, with the monks demonstrating a willingness to engage in political affairs in both countries.

Donald Eugene Smith (1965) underlines 'the advantageous position' monks have to participate in different forms of political engagement, and emphasizes their social prestige, network capabilities (both in urgency and reach) and their lack of mundane vulnerability (of losing their jobs, families or possessions). Another feat mentioned by Smith is what I would like to call a form of 'secular resilience', in that the police, military of legal institutions would be reluctant, even though they are not outright hindered, to bring various actions upon a monk. Gravers (2012) also notes how certain monks or laymen acquire status of 'enchanted subjects' that gives them a special form of spiritual and moral capital. However, the dynamics of such 'secular resilience' enables the monks to do various kinds of political oppositions usually barren to laypeople, and brings about certain reluctance from the regime to counter the activities of the political monks. This logic also leads to the somewhat confusing practice by the military regime in Myanmar of defrocking monks before interrogation.

When addressing the direct political methods employed by the various political monks in Myanmar and Sri Lanka, the practice of *pattam nikkujjana kamma*, or turning the bowl, would seem to carry the utmost potency in ways of resistance. While this practice also is known and sometimes used in Sri Lanka, it is to my knowledge never staged in a similar collective way of targeting a political entity. However, it should be noted that the title of the chapter by Juliane Schober dealing with the subject is named 'The Limits of Buddhist Moral Authority in the Secular State' (Schober 2011). Not even this massive event of de-legitimisation of the military regime was able to bring about its abrogation. While hunger strike was an effective means of protest in the colonial period in Burma, it has not gained momentum in present day Myanmar. Reports that monks commenced hunger strikes in detention in 2007 were followed by stories of the regime's defiance to this sort of protest ('we will beat you until you eat') (Schober 2011: 137). As such, the fast-unto-death campaigns of Ven. Omalpe Sobhita Thero in Sri Lanka have been far more successful, and in both instances some kind of action were taken in response to the protests. While monks (and others) supporting the independence of Tibet have committed self-immolation in a massive scale, only one instance of self-immolation has occurred in Myanmar in 2008. One method of political involvement that seems to have gone out of the repertoire in present times is the direct coercion by monks towards legislators, as the sit-in protest in Sri Lanka in 1957 and the coercive efforts in Burma in 1961, which shows that the methods employed by the political monks do adapt to their times.

A vital aspect to the dynamics of the political involvement of monks is the dynamics between the *sangha* as a unified hierarchy and the presence of temporary political formations to mobilize on particular issues. Monks have involved themselves in pressure groups, staged huge demonstrations and marches, and also formed their own political party, yet all of these efforts have been of a temporary nature. Neither in Myanmar nor in Sri Lanka have there been any attempts to create a solid organization for the promotion of religio-political engagement. The recruitment to these political formations predominantly comes from the younger parts of the clergy, and not the upper segment of the *sangha* hierarchy (Gravers 2012). Usually, they are as observed by Phadnis: 'loosely structured, ad hoc in nature and highly personalized in character' (Phadnis 1976: 273). Kovan (2012) argues that the frequency of the political interventions by monks, particularly by conducting alms-boycotts, do have an impact on the long-term political potential of Buddhist monks.

In times of political stability, the monks have the backing of the *sangha* as an institution to promote their interests, yet with the identification of a crisis, more radical means of resistance are allowed. Phadnis, on the other hand, argued that the *sangha*'s influence into politics can be formal as well as informal. Face-to-face communication with important decision makers, the issuing of resolutions and/or pamphlets could be just as effective as other forms of activist means (Phadnis 1976). The tacit agreement between the *sangha* hierarchy and the temporal political formations enables these dynamic influential lines of religious intervention into the domain of the political. My argument is that this tacit agreement between the *sangha* hierarchy and the temporary political formations offer its own dynamic; while the *sangha* as a whole is responsible for the societal influence (also in settings of politics), the temporary political formations field a more radical and activist-based form of pressure *in emergency situations*. The frequency and timing of such interventions by monks play a vital role in assessing the political potential of their activism.

Bibliography

Abeysekara, A. 2002. *The Colors of the Robe: Religion, Identity and Difference.* The University of South Carolina Press, South Carolina, SC, USA.

Amunugama, S. 1991. Buddhaputra and bhumiputra? Dilemmas of Modern Sinhala Buddhist monks in relation to ethnic and political conflict. *Religion* 21(2): 115-139. Available at: http://dx.doi.org/10.1016/0048-721X(91)90059-Y.

Bastin, R. 2010. Sri Lanka Civil Society and Its Fanatics. In: Kapferer, B., Telle, K. & Eriksen, A. (eds.). *Contemporary Religiosities: Emergent Socialities and the Post-Nation-State.* Berghahn Books, New York, NY, USA.

Deegalle, M. 2006. JHU politics for peace and a righteous state. In: Deegalle, M. (ed.). *Buddhism, Conflict and Violence in Modern Sri Lanka.* Routledge, New York, NY, USA.

DeVotta, N. & Stone, J. 2008. Jathika Hela Urumaya and Ethno-religious Politics in Sri Lanka. In: *Pacific Affairs* 81(1): 31-51.

Frydenlund, I. 2011. *Canonical Ambiguity and differential practices: Buddhist monks in wartime Sri Lanka.* University of Oslo, Oslo, Norway.

Gravers, M. 1993. *Nationalism as Political Paranoia in Burma.* NIAS, Copenhagen, Denmark.

Gravers, M. 2012. Monks, morality and military. The struggle for moral power in Burma-and Buddhism's uneasy relation with lay power. *Contemporary Buddhism: An Interdisciplinary Journal* 13(1): 1-33.

Imtiyaz, A.R.M. 2010. The politicization of Buddhism and electoral politics in Sri Lanka. In: Riaz, A. (ed.). *Religion and Politics in South Asia.* Routledge, New York, NY, USA.

Jordt, I. 2007. *Burma's Mass Lay Meditation Movement.* Ohio University Press, Athens, OH, USA.

Juergensmeyer, M. 2008. *Global Rebellion: Religious Challenges to the Secular State, from Christian Militias to al Qaeda.* University of California Press, Los Angeles, CA, USA.

Kovan, M. 2012. The Burmese Alms-Boycott: Theory and Practice of the *Pattanikujjana* in Buddhist Non-Violent Resistance. *Journal of Buddhist Ethics* 19.

Larkin, E. 2011. *Finding George Orwell in Burma.* Granta, London, UK.

Lintner, B. 2009. *The Resistance of the Monks: Buddhism and Activism in Burma.* Human Rights Watch, New York, NY USA.

Matthews, B. 1993. Buddhism under a Military Regime: The Iron Heel in Burma. *Asian Survey* 33(4): 408-423.

Owens, A. 2006. Using Legislation to Protect Against Unethical Conversions in Sri Lanka. *Journal of Law and Religion* 22(2): 323-351.

Phadnis, U. 1976. *Religion and Politics in Sri Lanka.* C. Hurst, London, UK.

Rahula, W. 2006. *The Heritage of the Bhikkhu.* Grove Press, New York, NY, USA.

Schmitt, C. 2007. *The Concept of the Political.* The University of Chicago Press, Chicago, IL, USA.

Schober, J. 2011. *Modern Buddhist Conjunctures in Myanmar.* University of Hawai'i Press, Honolulu, HI, USA.

Seneviratne, H.L. 1999. *The Work of Kings: The New Buddhism in Sri Lanka.* The University of Chicago Press, Chicago, IL, USA.

Shen, S. & Chan, P.C.-Y. 2010. Failure of the Saffron Revolution and the Aftermath: Revisiting the Transitologist Assumption. *Journal of Comparative Asian Development* 9(1): 31-57.

Silva, K.M.D. 1993. Religion and the state. In: Silva, K.M.D. (ed.). *Sri Lanka: Problems of Governance.* ICES, New Dehli, India.

Smith, D.E. 1965. *Religion and Politics in Burma.* Princeton University Press, New Jersey. NJ, USA.

Tambiah, S.J. 1992. *Buddhism Betrayed?* The Chicago University Press, Chicago, IL, USA.

Warnapala, W.A.W. 1978. *Sangha* and politics in Sri Lanka. Nature of the Continuing Controversy. *Indian Journal of Politics* XII.

Chapter 7: The second cultural revolution?

Individualization and the emergence of psychology in late modern China

Simen A. Øyen and Ole Jacob Madsen

The Forthcoming Empire

According to the popular grand narrative, supported for instance by the Communist party, China has undergone enormous changes since Deng Xiaoping implemented Reform and Opening Up [*Gaige kaifang*] in 1978, which introduced the reforms known as The Four Modernisations [*Si ge xiandaihua*].[102] The aim of this reform programme was to modernise agriculture, the armed forces, industry, and technology. Special economic zones were also established on the east coast, such as in Xiamen and Shenzhen. The establishment of private companies was permitted, as were foreign investments. In the aftermath of these reforms China has undergone a political and further economic liberalisation that, in combination with an authoritarian, centralised political power, has been called 'the new authoritarianism' (He 1997, Kokubun 1998). These socioeconomic and historical transformation processes have acquired cultural expressions and induced deep structural changes in terms of the ideals and the pressures on selfhood. This development is comparable to previously observed development tendencies in the West. However, it is not only the society at large and the economy as such that are undergoing changes; also the average Chinese citizen and Chinese subjectivity are being transformed at a rapid pace. Here individualisation processes, urbanisation, and the emergence of a consumer economy in particular are key developmental features. New lifestyles and changes in normative attitudes have arisen in parallel with Chinese modernisation and especially through individualisation and urbanisation processes. The Chinese term that was used to describe these institutional changes in the 1980s is *songbang* – a verb which literally means to 'untie' (Kleinman *et al.* 2011). These tendencies must be viewed in the context of the emergence of a civil society – complete with volunteer organisations and spontaneously organised enterprises alongside the state structures. Here, movements related to the environment and sustainable development, food safety and animal husbandry are particularly illustrative of these societal changes (see also Chapter 10).

[102] It is important to note that the Chinese economy was growing before 1978 and that the economic liberalisation is one of several hypotheses explaining China´s growth.

C. Brautaset and C.M. Dent (eds.) **THE GREAT DIVERSITY**
DOI 10.3920/978-90-8686-786-8_7, © Wageningen Academic Publishers 2014

The development of environmental movements that arose as spontaneous protests are illustrative of how individual political expression now can be articulated in a manner that was inconceivable 30 years ago. But we also find these tendencies in other areas of the civil society. China today offers many more opportunities for religious worship and the pursuit of private interests than was the case under Mao. Moreover, there is a close link between economic development in China and societal development in general and the formation of social movements, protest movements, and socio-political heterodoxies in the wake of the 1978 reforms in particular. This development has also comprised a transition from a period of industrial modernisation characterised by rapid industrialisation and material growth, to late modernity distinguished with post-materialistic values such as personal autonomy, life- and identity politics and rights assertion for women and marginalised groups. This development also illustrates a tension between endogenous factors within China and transnational and exogenous factors. These tensions involve not just issues concerning the actual development of an individualised consumer society in China in which professional expertise is gradually supplanting the function of family and kinships groups, between traditional Chinese medicine and Western medicine, Western systems of classification such as the DSM-IV and traditional concepts of mental illness. It also concerns a more general tension between what is perceived as Chinese culture and identity and the foreign and especially Western influence. The notion of 'individualisation' (Beck & Beck-Gernsheim 2002) has been vital in our understanding of how Western societies have developed from pre-modern states to modernity and finally into late modernity or post-modernity. Yet, it remains a highly contested concept (Beck *et al.* 1994). Our contribution to the ongoing debate is to discuss its transferability to experiences in China over the past decades. In the following we will look at some of the ways in which these developmental features find expression through shifts in the ideals of Chinese subjectivity. This book-chapter aims to present and expand on the research literature on individualization in contemporary China; especially the research based on the theoretical framework of the German sociologist Ulrich Beck. Here the Chinese social anthropologist Yunxiang Yan (2009, 2010, 2011) has played a defining and important role.

The British sociologist Nikolas Rose (2011) makes the argument that China goes through a psychological change of subjectivity and authority in light of the social changes of recent years. These entail conventional features we recognise from the West, but the psychologisation of subjectivity is also tied up with new, specific requirements to the late Chinese modernity. In the following we will address in further detail some of these developmental features and look at how diagnoses and therapeutic solutions have become the response to the problems experienced in the transition to a late capitalist consumer economy. Suicide and depression are now being redefined and diagnosed on a large scale, a commercialised self-help culture is on the rise, and new understandings and constructions of the subject are emerging as remarkable symptoms and expressions of cultural change processes and sociological individualisation processes also witnessed in the West. These phenomena are in different ways illustrative of increased individualisation. The structural changes in interaction between the individual and society also provide social scientists with a unique possibility to study comparative individualization processes that we in the West have come to take for granted.

Methodological framework

This book chapter is designed as an outline of the research status of recent sociological studies of individualisation processes in China (Yan, 2010) that typically investigate larger structural changes in the country's recent history by drawing on literature both from China studies and Western society. This allows us to test out conceptualisations like 'individualisation', 'enterprise self' and 'therapeutic culture' on China; concepts originally developed within a Western cultural framework. The selection of literature now available on recent social changes in China exceeds the sources we are able to present here. Thus, we do not aspire to present an exhaustive literary overview, but rather aim at giving the interested reader an small insight into some essential conflicts and contradictions both riding China and the recent attempts from Western researchers at capturing what goes on.

Theoretical framework

Beck and Beck-Gernsheim (2002) describe *individualisation* as an historical and structural process in which the individual to an increasing extent acquires a greater social significance. Individualisation processes must be viewed in the context of significant structural changes in the transition to modernity such as the industrial revolution, urbanisation, social mobility, the breakdown of the extended family, and women's rights. Beck & Beck-Gernsheim (2002) emphasise that Western society in late modernity also reproduces a risk in that the individualised social model advises us to find biographical and psychological solutions to systemic problems. In his classical work *Risk Society*, Beck (1992) applies the example of unemployment in West Germany during the first part of the 1980s. Here, Beck finds a tendency to hold individual employees personally responsible for the difficulties in which they find themselves due to career or vocational choices that were poorly adapted to future developments. The increasing forms of inequality produced by individualisation under late capitalism are the source of a common crisis for a great many people, but individualisation's fragmentation of personal destinies prevents employees from seeing this as a collective problem. Beck & Beck-Gernsheim (2002) cite the example of the development in China as a current demonstration of a prelude to individualisation. They describe the ongoing phasing out of traditional safety nets in China such as state subsidised housing programmes, education, and health services with an eye towards a new daily life in which individuals will be expected to personally take control over their lives. Another factor of the more recent individualisation processes emerges when identity evolves from being something that is defined by traditions and social and culture structures to a task to be mastered (Bauman 2001). It is therefore not unusual to speak of a turn in the West towards 'a therapeutic ethos' which is characterised by a change in the conditions of authority in which the self is given precedence over external laws and regulations (Lears 1983). Society is no longer something to be adapted to – but rather something that the self ideally is to be liberated from (Nolan 1998).

The birth of the modern Chinese individual

The Chinese Civil War (1946-1950) with the establishment of The People Republic of China in 1949 gave impetus towards liberating the individual from family and kinship restrictions; and also from the traditional social hierarchy (Kleinman *et al.* 2011). The individual was instead situated in an all-encompassing and pervasive socialist system of redistribution and political control. As a result, the individual identity was not solely formed through family and kinship groups, but was also defined by collective belongings, such as being a member of the rural communes [*gongshe*] or of the urban work units [*danwei*]. Following Deng Xiaoping's reform work in 1978, the people's communes and work units were phased out and the Chinese individual has gradually been released from collective and traditional structures (Saich 2004). The Chinese individual today therefore faces a '…fluid labor market, flexible employment, increasing personal risks, a culture of intimacy and self-expression, and a greater emphasis on individuality and self-reliance' (Yan 2009: 290).

As in the West, the emerging consumer economy has been a central driver behind Chinese individualisation processes. Accordingly, there is a close connection between the market and the development of civil society (Rofel 2007, Yan 2010). This is also the basis for a social mobility that was entirely inconceivable 30 years ago. Combined, these developmental features have provided the basis for the emergence of a new Chinese subjectivity in striking contrast to 'Chairman Mao's good socialist soldier' who was first and foremost loyal to the party state and put the important revolutionary ideas ahead of any individual interests:

> Market competition and social mobility encourage and even force the individual to be proactive, rationally calculating about self-interest, and competitive, which has led to the rise of what Nikolas Rose (1996) calls the 'enterprising self.' Since the mid-1990s ...the image of an enterprising self is commonly shared by Chinese youth who not only actively participate in various forms of self-development but also perceive the social world to be inhabited by autonomous and responsible individual actors. (Kleinman *et al.* 2011: 4).

The intensified social mobility and individualisation is an indicator of the growing Chinese middle class; a social class that did not exist under Mao (Kleinman *et al.* 2011). These tendencies are also generally speaking more pervasive for the Chinese society. For example, more recent studies of migrant workers – especially young people from rural areas who are working in the big cities on the east coast such as Shanghai, Shenzhen, Xiamen – show that they no longer leave the villages exclusively due to economic difficulties or to support and help the family economically, but also in order to realise themselves and take advantage of modern opportunities (Hansen & Thøgersen 2008).

The market has, as in the West, functioned as a channel for individualisation of the Chinese citizen after the death of Mao in 1976; albeit under the direction of the party state, which has succeeded in implementing a series of institutional changes such as the adaptations for increased mobility in the labour market (Yan 2009). The party state has contributed to these changes by implementing directives and institutional changes such as relaxation of the household registration system [*hukou*] and dismissal of workers from the state-owned enterprises. They also intensified the individualisation processes through disciplinary governing strategies. The most central strategy here was how the party state applied economic or political incentives in order to prevail by self-control rather than authoritarian rule and surveillance (Yan 2009, 2010). The party state has carried this out by promoting individual latitude in economic life, in private life, and in some select areas of the public sector. Market competition and social mobility encourage the individual to be more pragmatic with respect to its own interests. At the end of the 1980s and the beginning of the 1990s, the party state guided university graduates through teaching programmes and new regulations to enable and encourage them to apply for jobs in the labour market rather than being assigned a job by the state (Kleinman *et al.* 2011). Correspondingly, the three most important reform projects at the end of the 1990s – the privatisation of property, and the market adaptation of education and medical treatment – were institutional changes launched by the state to compel the individual to take more responsibility and become more actively involved in market-based competition (Kleinman *et al.* 2011).[103]

This can also be viewed as a governing strategy. The expansion of commercialised social spheres and consumer-generated individual autonomy redirects people's attention away from politically sensitive questions and towards questions about material well-being; something which to a large extent has helped the Chinese Communist Party (CCP) to maintain political stability and legitimacy after 1989. The political and economic liberalisation in China can therefore be indicative of a change from a repressive execution of power to a disciplinary execution of power, and thus from hard to soft control. Especially from 2004 and up to the present day, the dominant form of execution of power has been disciplinary (Saich 2004) with a shift to softer forms of control. The new principles for such a government were submitted at the 16th CCP Congress in September 2004 and cover this soft model of government (Sæther 2006).

However, it is not only the liberal market forces that have caused this development. Paradoxically, also the state's intervention in family life during the period of collectivisation accelerated individualisation processes in China (Yan 2010). The Communist party's attempts to direct the individual's loyalty away from the family and over to the local collective and finally, the state, resulted in a reinforcement of individualism. By replacing the family structure with collective structures, CCP in practise created new opportunities for the individual (Yan 2009, 2011). These two parallel processes have interacted in the establishment of – as we shall see – a predominant consumption-oriented and utility-based individualism.

During last part of the 20th century, the foundation of authority became more adapted to the individual's need for self-realisation. An example here is the observations made by Yunxiang (Yan 2011) in connection with field work done in the summer of 2008, where he took part in a study circle in Beijing. In one neighbourhood 10-15 people met once a week and studied books on ethics and social etiquette that were used in kindergartens and elementary schools in the US. They read the texts out loud to one another, discussed, and solved the exercises the books addressed. According

[103] These reforms were a part of the overall liberalisation of the economy that started with the reform and opening policy in 1978, escalated with Deng Xiaoping's symbolic journey to southern China in 1992, and with varying intensity has continued up to the present day (Saich & Kaufman 2005). This development has taken place through reforms for decollectivisation, legislation and central regulation, legalisation of new forms of local practices, state subsidised national programmes, campaigns, etc. Examples are the National Health Conference in 1996 which led to the new legislation that went under the name of 'Regulations for Health reforms and development: A focus on health and prevention in the rural districts' or guidelines issued by the public authorities for needs-based regional health planning in 1999 (Bloom 2005).

to Yan, the group had an explicitly therapeutic function in helping the participants to guide one another in relation to a moral life conduct in an increasingly more individualised and socially fragmented China. With the tensions involved in the transition from a planned economy to a market economy, the Chinese individual appears to have been asking for moral guidance as there: '...has been a shift away from an authoritarian, collective ethics of responsibilities and self-sacrifice toward a new, optional, and individualistic ethics of rights and self-development' (Yan 2011: 40). This shift produced new needs and an increased demand for psychology and therapeutic solutions in a country where psychology just a few years back was viewed as an extension of Western imperialism.

Psychology's more recent history in China

Seen from a Western point of view, it is not surprising that a new professionalised expertise thus has emerged in the form of psychologists and psychiatrists. During the Mao-era, different symptoms of mental and emotional problems were classified by the term 'neurasthenia', i.e. a diagnosis similar to the term 'nervous breakdown' (Lee 2011). Psychiatric ailments were generally viewed as bourgeois remnants from the capitalist society and could accordingly be cured through a politically formative education (Yan 2010). Psychological ailments were considered something that belonged to the old society [*jiu shehui*] along with drug abuse and prostitution. All of these social pathologies were symptoms of capitalism according to the dominant rhetoric under Mao.[104] The belief was that the new society [*xin shehui*] established by CCP would do away with all psychological illness while those who were already mentally ill would become well through the transition to a socialist society (Jinhua & Kleinman 2011). In consequence, there were many who were assigned with the single psychiatric diagnosis of 'neurasthenia'. As we will see, suicide was not defined and related to the psyche and psychopathology.

With the emergence of market capitalism, consumer culture, individualisation processes, and urbanisation psychiatry and psychology are increasingly in demand (Lee 2011). More people are now being diagnosed according to the international diagnosis system DSM-IV (soon to be replaced by DSM-5) produced by the American Psychiatric Association. How the authorities address large-scale accidents and natural catastrophes has been called the most extensive public initiative for

[104] This remains in part the case today and as late as in the beginning of the 1990s HIV/AIDS were viewed as a foreign and capitalist disease that could not exist in China: 'The corrupted Western bourgeois lifestyle' was believed to have provided the impetus for that which was labelled as social evils, such as prostitution, drug use, and homosexuality, and which resulted in various diseases including HIV/AIDS. At the time, the Chinese government asserted that the absence of HIV/AIDS in socialist China constituted substantial evidence for the superiority of socialism over capitalism (Jinhua & Kleinman 2011: 240).

psychological health of recent years (Lee 2011). This is also emblematic of this shift. The biggest earthquake of the 20th century in Tangshan on 28 July 1976 killed 242,419 people according to the official statistics (other sources claim that the number of fatalities was up to three times more) (Lee 2011). The Chinese authorities refused all offers of assistance from the United Nations. Little is known, even today, what type of assistance measures were implemented. Moreover, the potential post traumatic disorders the victims of this earthquake suffered also remain esoteric. No form of psychological assistance whatsoever was offered. After the earthquake in Wenchuan in Sichuan, however, on 12 May 2008, where an estimated 240,000 people died, the Chinese authorities sent 171 professional psychological health workers into the disaster area. The authorities also offered other forms of psychological health services, and the department of health published the manual *A short and simple guide to help individuals suffering from serious psychological trauma following a large earthquake* that was distributed to the military, health personnel, and volunteers working in the catastrophe region (Lee 2011, Peng n.d.).

With the type of development that China has undergone in terms of individual responsibility and the appeal for self-realisation new problems ensue. It can appear as if Freud's (2002 [1930]) classic argument on the cost of modern society remains relevant: the more advanced the form of civilisation, the higher the incidence of psychological illness is to be expected. The migrant workers who do not have the possibility to find a wife in a country with an enormous surplus of men, homeless in the big Chinese cities, homosexuals living under parental pressure to marry, and all those who do not immediately master the new, neoliberal, turbo-capitalism have acquired with time a dominant therapeutic industry from which to seek help. 190 million Chinese, or 13.5% of the population is believed to suffer from psychological problems with anxiety, depression, and substance abuse as the more frequently occurring ailments, according to the most recent figures from the Chinese Department of Health.[105]

Depression

The cardinal ailment in the West as a result of the prominent position of the self in late modernity is depression according to the sociologist Alain Ehrenberg (2010). His contemporary diagnosis corresponds to a large extent with the predictions of the World Health Organisation (WHO) that depression will be the most prevalent disease in the world by the year 2030 (BBC 9th February 2009). Viewed from the perspective of society, depression has now reached epidemic proportions precisely

[105] See for example *China Population Statistics Yearbook 2008* (China Statistics Publication Office 2008).

because the subject is obliged to live with the illusion that everything is potentially possible (Ehrenberg 2010). Depression appears to have become late modernity's equivalent to the neurosis of the Victorian era. Of course to make the claim that depression has *de facto* become the archetypical disease of modernity remains controversial. For instance, Horwitz & Wakefield (2007) in their book *The Loss of Sadness: How Psychiatry Transformed Normal Sorrow into Depressive Disorder* point to changes in the DSM system that now classify normal reactions to emotional stirring life-events as depression, as the main cause of the presumed epidemic of depression. This tendency is also possible to detect in China as psychological illness was previously defined as 'neurasthenia', whereas now one labels the same phenomena as depression (Lee 2011). Depression was in effect a completely unknown psychological category and diagnosis in China up until the early 1990s (Lee 2011).[106] The terms for depression [*youyuzheng* or *yiyuzheng*] or depressed [*yiyu* or *youyu*] were seldom used. In parallel to the dramatic and fundamental changes that arose with the reform and opening policy, the perception of depression has changed as well. Here a study by Arthur Kleinman (1982) is particularly illustrative. In 1978, Kleinman diagnosed 87 of 100 patients with depression, patients who had formerly been diagnosed with 'neurasthenia'. This study is indicative for the emergence of Chinese psychiatry. However, this is also representative for how the development of diagnoses in China must be viewed precisely in connection with the economic and social development.

While it might seem that the Western DSM-system's classification of depression is suitable for implementation with relative facility in relation to Chinese society, there are also specific features of the disease that must be understood in the context of China's more recent history, such as in terms of the one-child policy. Since the introduction of the one-child policy in 1979 in order to address the problem of increasing population growth, two studies conducted on Chinese students from the first generation of adult only-children found a higher incidence of narcissism than among their Western peers (Kwan *et al.* 2010). However, the adolescent generation of 'little emperors' appears to have frequent difficulties when it comes to meet the demands of working life and expectations from the family. In recent years a number of accounts have surfaced about an increased incidence of depression among the first adult generation of only children. The English language newspaper *China Daily*

[106] Two national epidemiological studies carried out in 1982 and 1993 found that depression occurred with a frequency of less than 0.5% in the Chinese population. After DSM's definition of depression was revised in 2009, individuals who feel 'emotionally constricted, gloomy/dejected, negative, pessimistic, and [who find it] tough to get through the day' are defined as depressed: 'Based on these controversially slackened criteria, a person who reported tough days or feeling pessimistic for two weeks and a few other symptoms of depression could be considered as having clinical depression. ...This elevates China to one of the most depressed countries in the world today! (Lee 2011: 202).

recently reported on the phenomenon in the article 'Post 80-s feel the pressure' (Yu 2010). Recently, the results from a study conducted on Chinese students reported that 60% of those interviewed answered that they felt isolated, while 80% stated that they experienced social injustice (Cheng 2010). In another survey of young adults more than half report a feeling of great pressure from their surroundings, while employees with a very high level of education struggle daily with exhaustion, sleep and eating disorders, dejection and loneliness (Yu 2010). The extent to which these estimates are representative is somewhat uncertain but they do offer an indication of the new challenges facing the adolescent generation of Chinese today.

Suicide

A widely held supposition in Western suicidology is that there is a relatively strong connection between psychological illness – especially depression – and suicide (cf. Arsenault-Lapierre *et al.* 2004). In China the connection between depression and suicide has been far from equally obvious. Suicide has until recently not been defined and connected to psychological illness to any particular extent, but instead viewed in light of issues such as honour and losing face. Here the symbolism connected to food offers insights of relevance. The anthropologist Wu Fei (2011) reports a case in Jianli where an elderly man apparently hung himself because his son and daughter-in-law had hidden bread from him and instead given him vile and rancid food. Another suicide case that Wu describes is also from rural China. Here the mother of the household, Eryao, usually prepared meals for the entire family. One day the daughter-in-law did the cooking instead, and she made just enough for herself and her husband. When Eryao found out that there was no more food left for her, she killed herself.

In conjunction with a fieldwork project Wu interviewed a man about suicide. The man told of a number of cases, but was unwilling to speak about the suicide of his cousin, a person who had feasibly had a mild cognitive disability. The man referred to the cousin as a fool – a not uncommon term for psychologically disabled individuals in China – and he therefore did not consider him to be somebody who committed suicide. In other words, the man did not define this case as something that could provide any insights or be appropriate for Wu's field work:

> My cousin Chaoyuan was a 'fool'. ...He was more than twenty years old, but unable to work, and he spent his days playing so that he was readily recognized by neighbors as abnormal. He was not considered to be a full person, but rather a 'fool'. ...One day he repeated some meaningless words. Because he appeared so foolish, I slapped him in front of several people. He returned

> home and drank pesticide and died. His mother and brother were also fools. (Wu 2011: 229)

Suicide is often viewed as a type of action that can only be carried out by mentally sound human beings. As in the example of Eryao, who's depression appears to be linked with family problems, suicide is understood as being a rational response to such problems, while depression is one of the consequences of precisely the same problems. Suicide is therefore believed not to be directly caused by mental illness as the mentally ill are perceived not to be in capacity of committing suicide as a moral act. This is, however, now undergoing change in correspondence with the new economic and cultural development. Not only is the number of suicides on the rise, but the increase can also be directly connected to the socioeconomic changes that China has gone through since 1978 (Lee 2011). The understanding of suicide in itself has changed correspondingly and in China this change appears to a large extent to be based on the presumed connection between depression and suicide that is prevalent in Western countries (cf. Cheung *et al.* 2006; Phillips *et al.* 1999).

A therapeutic culture in China?

It has been possible to discern the emergence of 'a therapeutic culture' in China, understood here as the gradual transition over the past 30 years from a collectivist ethics to an individualist ethics in the aftermath of the developmental features described above (Yan 2009, see also Chapter 10). Some scholars have argued that the most important driving power in Chinese society is, as in the West, the need to live one's life as authentically as possible as the creator of one's own unique identity (Taylor 1992). Individualisation transforms, as argued by Bauman, human identity from something fixed and given into a task. In the context of the economic liberalisation in China, the government has developed indirect techniques for managing and controlling the self by without being responsible for them (Rose 1996). Subjects are made accountable where former social risks such as illness, unemployment, and poverty is seen not primarily as the responsibility of the state, but also within the realm of influence of the recently empowered individual. The problem of ruling the collective is redefined at the individual level as a problem of self-control. Where 30 years ago most Chinese citizens were ensured a minimum of education, health services, and material well-being – referred to as 'the iron bowl of rice' – now the Chinese state has liberated the individual and individualised the risk. Since the mid-1990s the self-image of an 'enterprising self' full of initiative has become more and more common among young Chinese who are seeking self-realisation in an ever-expanding consumer economy and who would also prefer to perceive themselves and the social world as inhabited by other autonomous and accountable individual actors (Yan 2011).

The historical development in China has also led to the creation of more opportunities for self-realisation (Brandtstädter 2003). An individualistic-therapeutic world view opens for disciplines such as psychology and psychotherapy and the self-help industry. This is not only because they offer a variety of models for the self, but also because they provide practical formulas for action offered by professionals in different contexts as an answer to the spectrum of individualised, existential problems. These disciplines have acquired greater legitimacy in China after the Reform and Opening Up policy because the individual is made accountable to a far greater extent and the need for expertise seems imperative in many cases. We therefore see today similar tendencies as in the West, where some people turn to telephone and internet-based counselling services, therapy, and psychological consultations instead of seeking help from the family and social network (Wright 2011). The caring function that the family, relatives, or local society previously performed has today to a larger degree been taken over by professionals. The variety of telephone services that have become popular – particularly in large cities such as Beijing, Shanghai and Nanjing – serve to illustrate this (Frammolino 2004). For example, a suicide help-line was established at the Huilongguan Hospital in Beijing in 2002, and since start-up 10 years ago the service is estimated to have assisted in 110,000 consultations (Zhao 2010). Other common problems are heartbreak, marital problems and problems related to financial worries, teenagers, parents and depression.

The public authorities support this development by stimulating mental health care through an increased investment in psychology research, more mental health institutions, and psychological counselling services in the school system.[107] At the same time, the market for self-help literature designed to help people to find happiness or the meaning of life in China, as in the West, is experiencing remarkable growth. Examples of such books are 'Acrobatics for the soul' [*xinling ticao*], 'A guidebook for the soul' [*xinling zhinan*], 'What is happiness?' [*xingfu shi shenme*], and 'A beautiful life without a goal' [*wu mudi de meihao shenghuo*]. Yu Dan is one author who has achieved success in the ever growing demand for self-help literature by combining ancient Chinese philosophy with Western popular psychology. She has written about the Chinese philosopher Confucius and Zhuangzi and sold millions of books in which these philosophers offer guidance for the urban middle class of a China that is becoming ever more individualised.

[107] See e.g. http://www.cma.org.cn/ensite or http://www.doh.gov.tw/EN2006/index_EN.aspx.

Arguably, one could understand the self-help culture that focuses on the individual as a result of the liberation of the modern subject. When identity changes from being something that is a given or prearranged into something that becomes a task to be mastered, a need arises for professional expertise that can assist, treat, and realise the modern self. In this context, an expert culture has emerged that shall also guide the Chinese towards a continually improved self. The main purpose of the Chinese self-help literature would appear to be that of offering the reader a formula for how to master the balance between external expectations (obligations) and inner needs (rights) in the most important arenas of the modern Chinese woman and man, such as the home, the workplace and leisure. Accordingly it could be argued that the Chinese self-help culture for the time being, and for want of something better, serves an important guiding function for the modern Chinese subject who is torn between two sets of authoritative conditions.

From the iron bowl of rice to the therapeutic iron cage

The new sociological studies which focus on individualisation process and the rise of a therapeutic culture in China have opened up an exciting field of research that holds relevance both for scholars interested in contemporary China, and illuminates current debates on Western conceptions like modernity and individualization. In our small contribution to this field of study, we have drawn heavily upon on Yunxiang Yan's work, but we have also explored more thoroughly individualization, psychology and a therapeutic ethos as closely connected. In this article we have attempted to illustrate how the individualisation processes that China is going through have both clear features in common with Western individualisation and some evident differences. As a consequence of the economic liberalisation, the increasing social and cultural diversity, and urbanisation, China is undergoing individualisation processes that to some extent correspond with what is found in Western experiences. Correspondingly, there has been a shift from a collective ethical perception that emphasised the importance of self-sacrifice and hard work towards a collective goal such as the building of the new socialist society, to an individual-centred ethic with an emphasis on self-realisation and personal well-being. The new autonomous Chinese subject that has emerged within a very short period of time exhibits the historically unique social and psychological change processes that we have only just begun to understand.

Epidemiologists argue that the development outlined above is linked with a strong increase in depression and suicide. An emergent therapeutic ethos encompasses far more than institutionalised and professionalised psychology and psychotherapy. It is in the process of becoming a more comprehensive collective perceptual horizon.

This finds expression not only in institutionalised and academic psychology or through popular cultural expressions such as self-help literature, health advice, and therapeutic services for hire. It includes the very construction of the subject. This is a tendency that we also see in China and is an understanding that originates from the socio-historical individualisation processes that have been intensified, first through the economic liberalisation of the early 1980s, and then through the expanding Chinese consumer economy in the 1990s. In consequence, processes with a social origin are defined in individualising and self-centred terms, and viewed in light of a perception of authenticity and self-development. It is no longer the case that autocratic power structures such as the party state exclusively influence and form passive subjects. Individuals are left to define their lives and identity projects on their own (Rose 1999). In response to the experience of failure, diagnoses, therapy, and help are given at the one end of the self-realisation scale, and self-development and refinement of the self at the other. It is here that we may come to find the emergence of a therapeutic ethos with characteristic Chinese features – especially represented by the new generation of 'little emperors'.

Bibliography

Arsenault-Lapierre, G., Kim, C. & Turecki, G. 2004. Psychiatric diagnoses in 3275 suicides: a meta-analysis. *BMC Psychiatry,* 4(1): 37-47.

Bauman, Z. 2001. *The individualized society*. Polity Press, Cambridge, UK.

BBC. 2009, 9th February, Depression looms as global crisis, *BBC NEWS*. Available at: http://news.bbc.co.uk/2/hi/8230549.stm.

Beck, U. 1992. *Risk society: towards a new modernity*. Sage, London, UK.

Beck, U. & Beck-Gernsheim, E. 2002. *Individualization: Institutionalized individualism and its social and political consequences*. Sage, London, UK.

Beck, U., Giddens, A. & Lash, S. 1994. *Reflexive Modernization: Politics, Tradition and Aesthetics in the Modern Social Order*. Polity Press, Cambridge, UK.

Bloom, G. 2005. China's Rural Health System in Transition: Toward Coherent Institutional Arrangements? In: Huang, Y., Saich, T. & Steinfeld, E. (eds.). *Financial Sector Reform in China*: 158-178. Harvard University Asia Centre, Cambridge, MA, USA.

Brandtstädter, S. 2003. With Elias in China. *Anthropological Theory,* 3(1): 87-105. Available at: http://dx.doi.org/10.1177/1463499603003001753.

Cheung, Y.B., Law, C.K., Chan, B., Liu, K.Y. & Yip, P.S. 2006. Suicidal ideation and suicidal attempts in a population-based study of Chinese people: risk attributable to hopelessness, depression, and social factors. *J Affect Disord,* 90(2-3): 193-199.

Cheng, Y. 2010. Psychological issues rising among students. *China Daily*. Available at: China Daily website: http://www.chinadaily.com.cn/china/2010-03/10/content_9563769.htm.

China Statistics Publication Office 2008. *China Population and Employment Statistics Yearbook 2008*. China Statistics Publication Office, Bejing, China.

Ehrenberg, A. 2010. *The Weariness of the Self: Diagnosing the History of Depression in the Contemporary Age*. McGill-Queen's University Press, Montreal, Canada.

Frammolino, R. 2004. China Discovers the Couch. *Los Angeles Times*. Available at: Los Angeles Times website: http://articles.latimes.com/2004/sep/17/world/fg-china17.

Freud, S. 2002 [2009]. *Civilization and Its Discontents*. Penguin, London, UK.

Hansen, M.H. & Thøgersen, S. 2008. *Kina – individ og samfunn*. Universitetsforlaget, Oslo, Norway.

He, B. 1997. *The democratic implications of civil society in China*. Palgrave, Macmillan, Basingstoke, UK.

Horwitz, A.V. & Wakefield, J.C. 2007. *The loss of sadness: How psychiatry transformed normal sorrow into depressive disorder.* Oxford University Press, Oxford, UK.

Jinhua, Y. & Kleinman, A. 2011. Stigma – HIV/AIDS, Mental illness and China's Nonpersons. In: Kleinman, A., Yan, Y., Jing, J., Sing, L. & Zhang, E. (eds.). *Deep China: the moral life of the person*: 237-262. University of California Press, Berkeley, CA, USA.

Kleinman, A. 1982. Neurasthenia and depression: A study of somatization and culture in China. *Culture, Medecine and Psychiatry*, 6(2): 117-189.

Kleinman, A., Yan, Y., Jing, J., Sing, L., & Zhang, E. 2011. *Deep China: the moral life of the person*. University of California Press, Berkeley, CA, USA.

Kokubun, R. 1998. The Essence of Contemporary Chinese Bureaucracy. In: Brodsgaard, K.E. & Strand, D. (eds.). *Reconstructing Twentieth-Century China: State Control, Sivil Society, and National Identity*. Clarendon Press, Oxford, UK.

Kwan, V.S.Y., Hui, C.-M. & McGee, J.A. 2010. What do we know about the Chinese self? Illustrations with self-esteem, self-efficacy, and self-enhancement. In: Bond, M.H. (ed.). *Oxford Handbook of Chinese PsychologyI*: 279-294. Oxford University Press, Oxford, UK.

Lears, T.J.J. 1983. From salvation to self-realization. Advertising and the therapeutic roots of the consumer culture, 1880-1930. In: Fox, R.W. & Lears, T.J.J. (eds.). *The Culture of Consumption: Critical Essays in American History 1880-1980*: 1-38. Pantheon Books, New York, NY, USA.

Lee, I. 2011. Depression – Coming of Age in China. In: Kleinman, A., Yan, Y., Jing, J., Sing, L. & Zhang, E. (eds.). *Deep China: the moral life of the person*: 177-212. University of California Press, Berkeley, CA, USA.

Nolan, J.L. 1998. *The therapeutic state: justifying government at century's end*. New York University Press, New York, NY, USA.

Peng, L. *A Different Path from Tangshan to Wenchuan: China's Participation in the International Disaster Reduction Regimes and its Domestic Impacts* http://www.ln.edu.hk/ihss/humanconf/pengpaper.pdf.

Phillips, M.R., Liu, H. & Zhang, Y. 1999. Suicide and social change in China. *Culture, Medecine and Psychiatry*, 23(1): 25-50.

Rofel, L. 2007. *Desiring China. Experiments in Neoliberalism, Sexuality, and Public Culture.* Duke University Press, Durham, NC, USA.

Rose, N. 1996. *Inventing our selves: Psychology, power and personhood.* Cambridge University Press, Cambridge, UK.

Rose, N. 1999. *Governing the soul: The shaping of the private self* (2nd Ed.). Free Association Books, London, UK.

Rose, N. 2011. Biological Citizenship and its Forms. In: Zhang, E. (ed.). *Governance of Life in Chinese Moral Experience.* Routledge, New York, NY, USA.

Saich, T. 2004. *Governance and Politics of China.* Palgrave Macmillan, New York, NY, USA.

Saich, T. & Kaufman, J. 2005. Financial Reform, Poverty, and the Impact on Reproductive Health Provision: Evidence from Three Rural Townships *Financial Sector Reform in China*: 178-204. Harvard University Asia Centre, Cambridge, MA, USA.

Sæther, E. 2006. *The polyphony of critical press discourse in Chinese media.* University of Oslo, Oslo, Norway.

Taylor, C. 1992. *The ethics of authenticity.* Harvard University Press, Cambridge, MA, USA.

Wright, K. 2011. *The Rise of The Therapeutic Society.* New Academia Publishing, Washington, DC, USA.

Wu, F. 2011. Suicide, a modern Problem in China. In: Kleinman, A., Yan, Y., Jing, J., Sing, L. & Zhang, E. (eds.). *Deep China: the moral life of the person*: 213-236. University of California Press, Berkeley, CA, USA.

Yan, Y. 2009. *The Individualization of Chinese Society*. Berg, Oxford, UK.

Yan, Y. 2010. The Chinese path to individualization. *British Journal Sociology*, 61(3): 489-512. Available at: http://dx.doi.org/10.1111/j.1468-4446.2010.01323.x.

Yan, Y. 2011. The Changing Moral Landscape. In A. Kleinman, Yan, Y., J. Jing, L. Sing & E. Zhang (eds.), *Deep China: the moral life of the person*: 36-77. University of California Press, Berkeley, CA, USA.

Yu, T. 2010. Post-80s feel the pressure. China Daily. Availabe at: China Daily website http://www.chinadaily.com.cn/business/2010-04/19/content_9745896_2.htm.

Zhao, Y. 2010. Suicide line struggling. China Daily. Availabe at: China Daily website: http://www.chinadaily.com.cn/cndy/2010-03/04/content_9534489.htm.

Chapter 8: National policy, sub-national trajectories

Development of local models in China's health reform process

Lewis Husain

China is in the process of building social provision systems. A major strand of this is building health insurance systems and in reforming the healthcare system more broadly in an attempt to increase access of the rural population to affordable healthcare services. This is an attempt to reverse much of the process of marketisation of care and collapse or neglect of health insurance schemes which took place starting in the late 1970s. A key strand of this reform effort is the New Cooperative Medical Scheme (NCMS), a rural health insurance scheme aimed at providing broad-based, but low level, population coverage. Development of the scheme was started in the early 2000s. Overall, the scheme is described as a 'pilot' [*shidian*] in Chinese policy discourse, and counties,[108] which are principally responsible for running the scheme, have considerable freedom in its development.

An overall discourse of 'piloting' encompasses multiple dynamics in scheme development, from managed experimentation to encouragement and tolerance of local innovation. This latter phenomenon responds to two main beliefs 'in and around government' (Kingdon 1984: 125): first, that China's size requires local adaptation of policy and, second, that grassroots 'experience' developed during the reform process can contribute to reform and allow progressive refinement of policy. This is a pragmatic solution to policy development, given sub-national diversity in starting conditions and in local developmental trajectories, as well as the impossibility of the central government adequately understanding the range of existing conditions across the country. This chapter examines the development of an innovative NCMS oversight system in Taoshan,[109] a relatively poor county in an eastern province, to ask to what extent such peripheral policy development can contribute to development of national policy.

[108]'County' is used as a catchall term for a Chinese administrative level, to include also cities [*shi*] and districts [*qu*], and without reference to level of urbanisation, etc. See Lam on the place and importance of counties in the Chinese government system (Lam 2010). Below counties are towns/townships, which lack many of the government departments and functions present at the county level.

[109]Names of places and people have been changed to preserve anonymity. In some cases, figures have been marginally altered in order to preserve anonymity. For the same reason, county-level policy documents and reports are given only indicative references. In all cases any changes have no effect on meaning.

C. Brautaset and C.M. Dent (eds.) **THE GREAT DIVERSITY**
DOI 10.3920/978-90-8686-786-8_8, © Wageningen Academic Publishers 2014

In 2008, Taoshan County set up what the county describes as an Independent Oversight Bureau (IOB) to oversee operation of the NCMS, and this was reported on favourably by the Ministry of Health (MOH). Creation of the Bureau was a local initiative, but responds to a larger, and commonly-recognised, deficiency in scheme oversight at the county level. The 2nd section provides background on the NCMS and the 'distinctive policy process' under which this is being developed. The 3rd section analyses management and oversight of the NCMS in order to situate Taoshan's reform. The 4th section describes Taoshan's reform: what the county did and why. The 5th section examines Taoshan's reform in relation to reform at larger scales in Chinese government and the extent to which this mechanism is having, or could have, supra-local applicability and/or impact. The 6th section concludes.

The New Cooperative Medical Scheme and China's health reforms: a distinctive policy process?

The New Cooperative Medical Scheme

In rural areas, transition from a collective agricultural economy to household-based farming, from the late 1970s and early 1980s, was accompanied in most places by the dissolution of the Cooperative Medical Scheme, China's much-discussed pre-reform rural health insurance system. By 2003, and the start of piloting of the 'New' CMS (the *N*CMS), around 80% of rural residents lacked any form of health insurance whatsoever (Wagstaff & Lindelow 2008), and individual spending accounted for around 60% of total health spending, up from around 20% in 1978 (Hu *et al.* 2008, Wagstaff *et al.* 2009c). In most places, rural providers have suffered from years of inadequate government investment, one consequence of which has been that dramatic rises in healthcare spending – of around 16% annually since the early 1990s, faster than growth in GDP and incomes (Wagstaff *et al.* 2009c, Yip & Hsiao 2009), and price growth generally (Bloom & Fang 2003) – have been largely passed on to patients.

Together, these two trends have led to a decrease in risk sharing and individualisation of the burden of health spending. A recent World Bank report puts the out-of-pocket expense of a single inpatient episode at around 60% of *per capita* consumption (Wagstaff *et al.* 2009b), while other studies estimate this to be even higher (Xu 2008, cited in Hu *et al.* 2008). National Health Survey data show high levels of missed hospitalisation and early discharge against medical advice due to high cost of care (Wagstaff & Lindelow 2008). Those seeking care risk impoverishment: China's health system is one of the worst in the world when judged on protection from financial risks (Liu *et al.* 2008), and health spending has substantially increased China's poverty headcount (Van Doorslaer *et al.* 2006). Starting in the 1990s, China has been experimenting with various components of rural health reform, including an attempt (largely unsuccessful) to revive the CMS (Jackson *et al.* 2005, Wagstaff *et al.* 2009c). This is the background to announcement of the *New* CMS in 2002-2003, as part of a larger shift in attention to rural issues (Christiansen & Zhang 2009) under the Hu-Wen government.

The NCMS is run at the county level and aims to provide basic insurance against serious illness. The scheme is funded by sub-national government (counties and provinces) and user contributions, with additional funds from central government in less rich areas. Funding is based on central transfers once local government contributions are in place, which provides an incentive for counties to sign up users (but also to misreport numbers in order to claim more funds from the centre). Until very recently, the scheme has been run on *per capita* funding levels that make it hard to reimburse more than a fraction of users' health spending. One result of this is a variety of measures designed to limit liability of the scheme and to discourage non-essential use. A number of common measures are use of thresholds for NCMS reimbursement (i.e. the scheme only pays out once spending exceeds a certain level), user co-payments, payment caps, limiting reimbursement to inpatient treatment (excluding outpatient), and use of positive lists of reimbursable procedures and drugs. One consequence of this is that users face any number of reductions and co-pay elements, and are reimbursed at levels lower than those advertised by the scheme (Zhang *et al.* 2011). Providers face constraints on their activity, including limits on drugs and procedures that can be reimbursed by the scheme. In recent years, NCMS funds have increased substantially and are now at levels that stand a chance of reducing the burden of healthcare spending of rural populations (Herd *et al.* 2010).

A distinctive policy process?

Recent scholarship on China's policy process has emphasised the importance of managed experimentation (Heilmann 2008a), and on capturing the gains from decentralised policy 'tinkering' (Heilmann & Perry 2011) in policy development. The use of local experiments in reform has received a degree of attention in the health literature for some time (Bloom *et al.* 2000), though there is very little attention to actual processes of experimentation in development of specific policies and how sub-national governments function in these processes.

The NCMS is described in policy discourse as a 'pilot' [*shidian*], but what this refers to is frequently not specified. In practice, it is possible to distinguish between, on the one hand, managed experiments carried out to test particular working methods in a relatively controlled manner and, on the other, Heilmann's decentralised 'tinkering'. Both modes of policy development are important in explaining the array of reform dynamics visible in a policy area such as the NCMS. One of the most widely discussed managed NCMS experiments, for example, is Lufeng County's [Yunnan] development of NCMS provider payment reforms, overseen by Wang Lusheng of the China National Health Development Research Centre (author's interviews with county NCMS staff, June 2010). Similarly, Qianjiang District [Chongqing] has been involved in multiple iterations of the China Rural Health Project supported by the Ministry of Health, World Bank and UK Department for International Development (Zhang *et al.* 2007). Equally, in 2008 the MOH announced experiments intended to develop specific facets of NCMS policy, and specified counties in which pilots were to be conducted, experts charged with overseeing progress, etc. (Nongweisi 2008). Clearly, counties involved in such pilots have received substantial capacity building and oversight and have a high degree of centrality in China's health system reforms.

Alongside this, though, non-pilot counties face an imperative to develop local policy solutions, and county innovation is both encouraged and tolerated by the centre. As Vice Premier Wu Yi explained at the 2007 national NCMS meeting, 'We use a method in which the centre sets the principles of policy and localities concretely organise implementable policy; we encourage each locality to experiment and innovate' (People's Daily Online, 16th March 2007).[110] Underlying this is a belief that local 'tinkering' is necessary, in order to allow counties to 'implement in line with local conditions' [*yindi zhiyi*] and that piloting or experimentation, whether explicitly guided or organic, will produce valuable 'experience' [*jingyan*] useful for reform.[111] The level at which this is expected to work is that of policy 'mechanisms'.[112]

Over 2002-2003, development of a rural health insurance framework was begun, starting with a small number of counties and small amounts of money and with the expressed aim of reducing impoverishment resulting from health seeking. Since then, this system has evolved very rapidly and has been rolled out nationwide. Ten years on, while funding and reimbursement rates have increased greatly, the scheme remains a work in progress. A number of challenges in scheme development remain.

[110] In line with convention, 'locality' is used to refer to the county level.

[111] This has a clear precedent in Mao's *On the ten major relationships* (Mao 1977), which has been core reading for successive generations of China's leadership (Zheng 2007). See also Heilmann's treatment of the origins of China's 'distinctive policy process' (Heilmann 2008b), and Schurmann's classic treatment of the 'regions' in policy adaptation (Schurmann 1966).

[112] My use of mechanism corresponds to '*jizhi*' in Chinese, which I conceive of as a component and dependent element of a larger, and separately instituted, policy.

Given a central commitment to local initiative and innovation, and given that most counties lack the resources and guidance that come with a centrally-supported pilot, to what extent can such places develop workable policy solutions, either tailoring policy to local conditions or contributing to policy development at larger scales? This chapter examines one county's experiment with development of a local NCMS oversight mechanism, an acknowledged deficiency in scheme design. This supplements existing analyses of differences in a narrow range of programme variables (mainly reimbursement rates, ceilings, deductibles, coverage of inpatient versus outpatient spending) aiming to evaluate scheme effectiveness (Brown *et al.* 2009, Wagstaff *et al.* 2009c) in looking at counties' development of policy solutions under an avowedly experimental policy process. The next section discusses NCMS management at the county level and the structural conjuncture limiting oversight of the scheme. This is background to Taoshan's reform, discussed subsequently.

New Cooperative Medical Scheme: under county management

New Cooperative Medical Scheme management and oversight

At the county level, and in the majority of cases, the NCMS is overseen by an NCMS Management Committee [*Guanli Weiyuan Hui*], with day-to-day management carried out by an NCMS Management Office [*He Guan Ban*, hereafter NCMS Office]. This structure may be replicated at the town/township level, though this is not universal (Gao & Han 2007). The NCMS Committee falls under the dual leadership of the county government and the Bureau of Health (BOH). Broadly speaking, the NCMS Office is charged with day-to-day oversight and management of the scheme (NCMS Pilot Evaluation Group 2006). Bread and butter work revolves around overseeing fund use, overseeing and processing claims (approving claims/ payments), and overseeing providers included in the scheme (Gao & Han 2007). Structurally speaking, the NCMS Office should perform a gate-keeping and oversight role within the county-level management of the scheme, standing at the interface between the finance administration [*xitong*], providers and users.

Two main problems exist in relation to the functioning of the NCMS Office in oversight of providers/services and the claims/reimbursement process: lack of funding, personnel and capacity; and lack of independence from the local health system. The capacity of NCMS Offices to carry out effective oversight and management in most places is limited. A considerable Chinese literature has discussed the limited staffing and capacity of NCMS Offices stemming from limited funding and personnel [*bianzhi*] allocations. Funds for scheme management must be found from county budgets, and can be considerable – including spending on staff salaries, office space, cars, computer systems, administrative fees, etc.[113] Especially in poorer areas, such expenses can be a very considerable burden on county finances (Gao *et al.* 2006). Spending on management of the NCMS was, at least in early stages, around a fifth of *per capita* management spending on the urban employees' insurance scheme [*Chengzhen Zhigong Yiliao Baoxian*] (Wang *et al.* 2007, cited in Gao 2008). An interim State Council mandated report on the scheme found that limited funds had an impact on implementation and on oversight and management work (NCMS Pilot Evaluation Group 2006).

The personnel allocation of most NCMS Offices is set at 5 to 6 people, meaning that, on a *per capita* basis, county-level NCMS managers are responsible for oversight of a population of around 50,000 people (Gao 2008). Such staffing levels are clearly inadequate, and NCMS Offices frequently rely on assistance from employees of hospitals and township health centres (THCs) under their purview to help oversee NCMS operation in those same providers. In practice, this frequently means that designated providers (those providers authorised to treat within the NCMS) must allocate staff to NCMS duties, including overseeing reimbursements and reporting on scheme operation. One review of scheme functioning found that half of township-level personnel were sent down by the county, with half locally seconded (NCMS Pilot Evaluation Group 2006). Staff seconded from providers, however, remain on the payroll of the provider in question, and continue to be managed by their 'parent' provider, despite performing duties for the NCMS Office, creating a clear possibility of conflicts of interest in oversight of providers. Interviews with NCMS managers in several counties confirm that staffing levels are inadequate, and that this creates problems for oversight of providers and scheme functioning overall.

Other conflicts of interest also exist. The position of the NCMS Office within the local health system is deemed to cause difficulties for effective oversight of providers: historically in China, local health administrations (here, the county BOH) and providers (here, principally county hospitals and township health centres) all fall within one government system [*xitong*], a vertical administration with the MOH at its head. In the case of the NCMS, this has potential to cause conflicts of interest as county bureaus of health oversee both the payer (NCMS) and payee (providers) components of the local health matrix. Concretely, in most cases, the BOH appoints the head of the NCMS Office, as well as having oversight of personnel, finances and material supplies of providers (NCMS Pilot Evaluation Group 2006). Given historical government under-funding of providers and limited responsiveness of

[113] Following debates in the 1990s and onwards about the 'peasant burden' [*nongmin fudan*], when the NCMS was launched, it was specified that management expenses not be taken from scheme funds, which are to be used exclusively for reimbursement (Guobanfa 2003, no. 3).

local governments in many places to local populations, there is a fear that bureaus of health see their interests as coinciding with those of healthcare providers rather than NCMS users (Yip & Hsiao 2008).

Limited capacity available for management of the NCMS is to a large extent a product of hurried introduction of the scheme, with little attention to practicalities of management or where funds were to come from for management (Yang 2004), as well as the specific conjuncture surrounding launching of the scheme. Following debates in the 1990s and onwards about the 'peasant burden', when the NCMS was launched, a new rural health insurance scheme – while deemed necessary – could not be seen to increase the peasant burden at a time when government was working to eliminate this (Zhongfa 2002), and it was specified that management expenses not be taken from scheme funds, which are to be used exclusively for reimbursement (Guobanfa 2003).

County-level oversight

Counties implementing the NCMS should set up Oversight Committees [*Jiandu Weiyuanhui*] (Guobanfa 2003), whose function is to check whether NCMS funds are in place and are being spent and managed correctly (specifically managed in line with Ministry of Finance regulations), to check whether balance is being maintained in fund use (between revenues and payments), to check on implementation of NCMS spending plans, to publicise information on the state of NCMS fund use and investigate incorrect and illegal behaviours (NCMS Pilot Evaluation Group 2006). In theory, the Oversight Committee should ensure correct scheme operation, but this is disputed by many analysts of the NCMS. Two main criticisms stand out. The first is the lack of technical expertise of Committee members, who are unlikely to have medical training, making oversight of limited value. Second, Committees tend to be staffed by People's Congress and People's Consultative Conference members, and to have close links to local government. Two interim reports on development of the NCMS, commissioned by the State Council, both come to the same conclusion: that oversight by these committees is ineffectual (NCMS Pilot Evaluation Group 2006, MOH Center for Statistics Information 2007). Interviews with county NCMS managers confirm the view that local People's Congresses lack the ability to carry out effective oversight (author's interviews with county NCMS staff, May 2010).[114]

[114]In addition, other systemically less significant factors, such as the fact that Oversight Committees inevitably carry out after-the-fact oversight, rather than real-time oversight, are also likely to impair their effectiveness (Chen 2010). In addition, it is thought that Committee members tend to be at the end of their careers and unlikely to have an incentive to carry out effective oversight.

NCMS regulations specify that local audit departments should carry out periodic audits of scheme funding (Guobanfa 2003), and that NCMS Committees and NCMS Oversight Committees should include user representatives [*nongmin daibiao*]. Interim reports show high numbers of both committees to have user representatives (NCMS Pilot Evaluation Group 2006), but available analyses point to a lack of real importance of this participation (NCMS Pilot Evaluation Group 2006, Gao 2008, Chen 2010). Many analyses point to various problems as symptoms of poor oversight at the county level. While serious scheme financial mismanagement is believed to be rare due to the existence of solid finance and accounting systems, it is believed that many problems exist in oversight of providers and NCMS reimbursement. Examples frequently cited are illegitimate claims against the scheme and users illegitimately acquiring prescriptions for others, over-provision of treatment, doctors (and/or doctors in collaboration with patients) submitting false claims in order to cheat the scheme and similar. Lack of effective oversight of providers is believed to contribute to inflation of diagnoses and unwarranted cost increases (NCMS Pilot Evaluation Group 2006, Gao 2008).

Any number of other management problems is likely to exist, to different degrees, in different places. Different reports present different worries, though there is a degree of consistency: lack of implementation financial management systems, use of incorrect reimbursement practices, etc. Studies have even found certain counties to be using NCMS monies to pay for users to sign up to commercial insurance (NCMS Pilot Evaluation Group 2006, Wu 2009). Summary results of a 2009 National Audit Office audit of the NCMS state that no large problems in breach of regulations were discovered, but that numerous smaller problems were discovered, including misuse of funds (including routing NCMS funds into other social security programmes), incorrect reporting of numbers of scheme members with the intention of claiming higher levels of central matching funds, insufficiently rigorous oversight of reimbursement resulting in wasted monies, etc. (Zhonghua Renmin Gongheguo Shenjishu Bangongting 2011).

County-level NCMS oversight is, in most places, to be considered weak. This is a systemic weakness arising from funding, capacity and institutional constraints. While most analyses believe that major problems of fund diversion should be considered rare, a variety of common problems are highlighted by analysts of the NCMS and these have the potential to undermine scheme functioning and to increase costs. For a scheme operating for most of its existence on low budgets, this is obviously a worry and could prejudice sustainability of the scheme. The following section examines Taoshan's initiative: what the county did and the thinking behind this.

Taoshan's local oversight model

From 2008 to 2009, Taoshan trialled its Oversight Bureau in two townships, before rolling this out county-wide. The BOH approached county government for permission (and funding) to carry out a local pilot, and this was given. Local piloting allowed the BOH to trial the system before roll out, convince local government of its usefulness in improving scheme operation and user satisfaction, and to minimise risk. Taoshan's initiative has received a degree of attention outside the county, including from the MOH and has been studied by other counties, but has not achieved significant national fame (see 5th section). Taoshan's Oversight Bureau aims to separate scheme oversight and management: oversight is placed in the Oversight Bureau, not with the NCMS Office, and the Bureau is independent, financially and in management terms, from the NCMS Office and from providers, reporting to the head of the BOH and the NCMS Committee.

Factors underlying development of Taoshan's Oversight Bureau

Both local and systemic factors underlie development of Taoshan's Oversight Bureau. According to analyses published by the BOH, and to interviews with BOH staff, a major factor underlying development of the IOB is the increasing funds under NCMS management. National NCMS funding levels increased from 30 RMB in 2003 to 100 RMB in 2008 (and more subsequently) (Weinongweifa 2008). Prior to establishment of the IOB, Taoshan encountered various problems in NCMS oversight, which the county describes in terms of oversight of NCMS users, providers and NCMS implementation agencies. Problems listed in BOH reports include excessive drug prescription, inappropriate use of drugs, inappropriate reimbursements, high costs per case due to lax monitoring, over-prescription and similar. NCMS implementation agencies are described as being lax in their work. None of these problems should be seen as special or unique to Taoshan; rather, they are common to hundreds of NCMS policy documents, and presumably to hundreds, if not thousands, of counties around China.

The diagnosis by the county of the reason for these problems is very close to the diagnosis given above of the scheme more broadly: the NCMS Office (and Committee) faces difficulties in overseeing the NCMS, there are too few staff to carry out effective oversight, and the model of oversight is faulty, relying on staff drawn from THCs, which creates a conflict of interest (Taoshan County Bureau of Health 2010). Taoshan BOH staff describe the system of relying on THC staff to carry out oversight of those same THCs as 'self-monitoring'. As above, given low levels of government support for providers and THCs' reliance on earned income, Taoshan BOH staff argue that this model of oversight can only be ineffective as THCs will inevitably place their own revenue generation and survival before the interests of the NCMS scheme. This inevitably damages implementation of the policy and reduces its effectiveness. In some cases, they claim, providers' internal oversight exists 'in name only'. This is compounded by inadequate staffing levels: Taoshan's NCMS Office oversees almost 400 providers and around 900,000 users. This is to say that, as above, Taoshan's initiative responds to a problem – and framing – general to the NCMS, rather than uniquely local problems, though there was likely a significant local catalyst. The aim of the Bureau is principally to strengthen 'external management' of providers in order to make up for their insufficient 'internal management'.

Work of the Bureau

The main work of the Oversight Bureau is as follows: to supervise and encourage township-level NCMS fundraising; to check on NCMS reimbursements, making sure that they are within the scope specified by the NCMS; to check that township-level NCMS Offices are effectively monitoring village-level NCMS reimbursements; to check fund collection and use at village, township and county levels; to collect public complaints; to check injury claims;[115] to assist providers in carrying out internal auditing; to publicise the NCMS; to report on oversight work; and to carry out door-to-door surveying. This work is carried out by the Bureau itself, housed within the BOH, and through five Sub-Bureaus, each of which oversees a 'district'[116] within the county, and is staffed by a manager and three to four other staff. Core work is carried out by Sub-Bureaus, whose work is supervised by the county-level Bureau (Taoshan County Bureau of Health 2009b), and the work of Sub-Bureaus is to be assessed according to standardised management procedures, with results linked to bonuses and other benefits (Taoshan County Bureau of Health 2009a) as in government work more generally.

In practice, various kinds of oversight are carried out by the Bureau and Sub-Bureaus. The majority is routine inspection of accounts and patients' medical files (records of treatments). In such cases, providers are notified in advance and this functions as an audit of their treatment and spending under the NCMS. Unannounced and targeted inspections of providers are also used in cases where illegitimate behaviour is suspected, or where it is thought that prior notification might prejudice inspection.

[115]The NCMS frequently does not pay where there is third party liability, notably in injury cases. Determining whether a third party is liable frequently requires significant checking on the part of NCMS implementation agencies.

[116]Management is based, here, on a spatial unit below the county level, but not corresponding to the existing town/township administrative structure. This is routine in some areas of administration, including police and traffic administration, but not routinely used in the health system.

The Charter of the Bureau specifies in detail procedures for inspection, with specific, standardised, documentation requirements for each of stage of the inspection process: from documents notifying targets of forthcoming inspection and forms for requisitioning providers' records, to forms detailing possible decisions by the inspection team. Possible outcomes of cases include positive judgements and consequent closing of a case file, requests that a provider rectify practices within a specified time period, and orders for change and notification of impending re-inspection to check compliance.

Infractions discovered by the Bureau or Sub-Bureau are to be dealt with initially through a request for change, though providers can be fined, if necessary, on the authority of the BOH. In extreme circumstances, providers' right to receive reimbursement from the NCMS can be suspended or removed, disbarring them from the NCMS system. Infractions by providers are listed in order of severity, and stretch from the comparatively less serious (refusing patients, incorrect referral, treatment or drug use, non-observance of NCMS pricing policy) to definitely serious (creating fake treatment records, fake drug prescriptions, excessive drug prescription, conspiring with patients to cheat the NCMS and similar).

Infractions by NCMS implementing agencies follow a similar pattern of minor and serious. Minor infractions, necessitating a warning, include the keeping of unclear accounts and the like. Serious infractions, liable, at least in theory, to a fine of up to 50,000 RMB or referral to legal departments, include: insufficiently rigorous checking of reimbursement documents; ignoring or disguising problems discovered during oversight of providers; not properly investigating injury cases; incorrect management of village-level reimbursement; deliberately cheating the NCMS. Possible infractions by NCMS users cover such things as use of another person's NCMS card to seek treatment, deliberate misrepresentation of illness, submission of false receipts for reimbursement and selling on of drugs prescribed through the NCMS for profit.

While the Bureau has oversight of three separate components of the health system, oversight of providers is clearly considered the most important and common. Bureau and BOH staff stress that most cases are resolved amicably, with providers agreeing to comply with Bureau requirements. The rationale for this is that while deficiencies in providers' 'internal' management create a need for increased 'external' management, the aim of external management should be to encourage providers to improve their own, internal, management practices.

Available case records from inspections carried out in early 2010 broadly show the working of the Bureau and Sub-Bureaus. Inspections of village clinics uncovered various kinds of incorrect behaviours, including: incorrect reimbursement procedures; excessive drug prescriptions; falsifying of prescription receipts by village clinics; discrepancies between records kept by village clinics and drugs actually prescribed, uncovered through door-to-door investigations; use of users' NCMS cards by village clinics to fraudulently claim expenses from the NCMS scheme, etc.[117] The severity of infractions varies, but Bureau inspection records provided during fieldwork and interviews indicate that these kinds of infractions are broadly representative of the kinds of problems the Bureau must deal with. Likewise, judgements regarding these cases vary, with the Bureau recommending change and publicising of problems (but not specifying punishment) in some cases, to fines and a recommendation in one case that the manager of a village clinic be sacked.

Available analyses are positive regarding the impact of the IOB. A May 2010 report from Taoshan BOH claims that in a year and a half of operation, the Bureau had carried out 1,800 inspections, and discovered 480 problems of various types. This report also credits the Bureau's work with helping reduce costs, increasing the proportion of reimbursable drugs prescribed,[118] and reducing spending on drugs overall.[119] The Bureau is also credited with improving performance of NCMS implementing agencies, and the % of reimbursements conforming to scheme guidelines is said to have increased from 95 to 99% over the period. Efficiency gains achieved by the Bureau are also credited with allowing the county to increase NCMS reimbursement rates to users (Taoshan County Bureau of Health 2010).

Taoshan's initiative as response to institutional constraint

Taoshan's initiative is very simple. Its core is the creation of an independent and resourced oversight agency, reporting directly to the NCMS Committee and the head of the BOH. It does this in two ways – one managerial and one structural. In management terms, the Bureau employs dedicated staff with expertise in a number of key areas (clinical medicine, accounting, computing or public health), who are placed in charge of dedicated offices each with responsibility for a number of townships under the county. This is a simple attempt to both increase the professional competence available for NCMS oversight and to locate this oversight much closer to the objects of oversight – the township and village level healthcare providers and

[117] Each user has an electronic NCMS card, which is used to access treatment/reimbursement.

[118] As above, not all drugs and treatments are reimbursed by the NCMS, and especially higher cost and less essential drugs tend to be excluded from the scope of reimbursement. This provides an incentive for providers to supply these drugs to patients outside the NCMS, requiring that they pay out of pocket. This has obvious implications for the ability of the scheme to increase users' financial protection.

[119] Spending on drugs is artificially high in China for structural reasons as providers sell excessive drugs to generate revenue (Eggleston *et al.* 2006: 1-2).

township level NCMS implementation agencies. The incentive structure and careers of Bureau staff are tied to performing oversight on behalf of the BOH.

Structurally, the Bureau attempts to avoid conflicts of interest locally through two main mechanisms. First, the existence of the Bureau in and of itself is an attempt to change local reporting lines and information flows. The Bureau exists in parallel to oversight work of the NCMS Office, and is under direct management of the head of the BOH and the NCMS Committee. This provides a direct reporting line from the township level straight to the NCMS Committee. Second, while Taoshan continues to use staff seconded from THCs, as in many other places, the Bureau attempts to resolve conflicts of interest at the THC level by changing the nature of the employment relationship of these seconded staff. These staff remain formally employed by the THCs in question, and money for their salaries continues to come from THC budgets, but these funds are routed through the Bureau, which pays salaries and controls management of staff (Taoshan County Bureau of Health 2008).

This mechanism allows the Bureau to effectively mobilise a staff of 25, while only putting its own core staff (the Bureau head, heads of Sub-Bureaus, inspectors) on the Bureau's payroll. In other words, its staffing costs are passed on, through government fiat, to the THCs. The main other cost is transportation expenses (one member of Bureau staff estimated spending around half of her time at the grassroots, doing oversight and door-to-door work). In this way, Taoshan achieves a degree of independence for its Inspection Bureau, isolating it from both the NCMS administration more generally and from the THCs (and subordinate village clinics) that it is charged with overseeing.

To say that this is a simple initiative is correct, but this risks obscuring the significance of what Taoshan has done. From a systemic point of view, the IOB responds to failings in the institutional design of the NCMS, which, for very specific and contingent historical and political reasons, concentrates oversight at the county level while giving insufficient attention and resources to this: as discussed above, oversight is hobbled by capacity, personnel and funding constraints and insufficient separation of management and supervision organs. The core of Taoshan's initiative is extremely simple: it strengthens local oversight of scheme functioning by requisitioning resources from THCs and placing these under the managerial control of the BOH and in this way makes good, at least to a degree, local capacity and management funding deficits. The following section examines Taoshan's initiative in the context of national policy development and asks to what extent it has had, or could have, supra-local significance.

Policy impact of Taoshan's Inspection Bureau

Taoshan's initiative in local and systemic context

As above, many analysts of the NCMS argue that oversight of the scheme is inadequate and should be strengthened and this is frequently asserted in central policy and speeches. To what extent does Taoshan's initiative provide a possible model of strengthened local oversight? The most significant calls for change argue for increasing third party oversight of the NCMS or increasing independence of oversight organs from local government. Such suggestions range from the simple, including use of independent accountants to provide third party oversight (Gao & Zhu 2009), to the more complex and systemically significant. Wu (2009), for example, argues for substantial reorganisation of the local health system and the separation of the health administration (BOH) from providers, and the development of a pluralised oversight system, whose guiding principle would be the independence of oversight from the local health system. Wu argues for the formation of professionalised groups at the provincial level, made up of people with relevant expertise from large hospitals, health research institutes, audit departments and so forth, to be selected at random to form a temporary oversight group. He also argues for solid transparency frameworks, setting out rights of access to information by People's Congresses and scheme users as well as greater media oversight. Some analysts go even further. Chen, for example, argues for greater user engagement in oversight and management of the NCMS, modelled on China's village committees (Chen 2010).

Taoshan's Oversight Bureau is described by the county as 'independent'. This 'independence', however, should be understood to indicate independence from healthcare providers and from the NCMS administration (especially lower levels), rather than independence from government or the BOH. Taoshan's Oversight Bureau clearly falls short of the above calls for systemic change, and is a reform of limited scope. It did not set out to increase transparency outside government, but rather to provide increased internal transparency. Director Wang, the head of Taoshan BOH, and the man behind the creation of the IOB, explained that he sees many reports describing problems in NCMS management, both in media and in internal documents. Most problems are not discovered from within the BOH or NCMS system, but by external audit committees and the like. Much better, according to this reasoning, to discover problems and correct them yourself than have others discover them for you.

To this extent, Taoshan's initiative arises from a sense of risk in implementing the NCMS. Director Wang described management of the NCMS before establishment of the Bureau as being similar to 'standing on the edge of a volcano': NCMS funds were substantial, and different parties in the health system had different interests but he had no way of knowing what was really going on or of maintaining control. Any problem in NCMS operation, he said, would be his responsibility; hence the need for the Oversight Bureau[120]. This risk framing is borne out by provincial records, which show substantial leadership turnover in Taoshan's health system in the immediate run up to the establishment of the Bureau, indicating a probable crisis in NCMS management as a proximate cause in creation of the Bureau.

At the risk of downplaying the crisis that Taoshan seems to have experienced, Taoshan's initiative should not be considered solely a local response to crisis, and should by no means be considered unique. First, the perception of risk underlying this kind of initiative is not unique to Taoshan. The head of the NCMS Office in Meijiang City, a prosperous, high capacity, county level city in the same province as Taoshan, also discussed NCMS implementation in terms of risk. Meijiang's response to this has been to contract day-to-day management of the scheme to a commercial insurance company, which can devote much greater resources to oversight than the NCMS Office would be able to do. Second, many other places in China have established similar agencies for similar purposes. Qianjiang [Chongqing] District's Accounting Centre, for example, performs a similar (though more extensive) function to that of Taoshan's Bureau (Zhang *et al.* 2007), and MOH policy briefs frequently carry mentions of such initiatives.

Third, Taoshan's initiative, as well as responding to a specific local conjuncture, also responds to widely-recognised deficiencies in scheme design, as above. Bureau staff argue that if counties do not strengthen NCMS oversight, they will lose money from the scheme as funds will be misspent. According to this logic, it would be reasonable to allow a small amount of money from NCMS funds to be used for this purpose as this would save the scheme money overall. This, however, would require national policy change and is unlikely to happen. Taoshan's initiative is a simple way of making good deficient national policy. In adopting this solution, what Taoshan has done is little more than formalise practices existing in many places and introduce a degree of managerial rigour into their functioning, if we consider that the co-opting of THC employees into running and overseeing the NCMS is, in fact, common in many places: Taoshan has simply formalised a *de facto* common administrative reallocation of resources in a way consistent with Chinese government personnel allocation [*bianzhi*] rules. Given this, how should we evaluate Taoshan's Oversight Bureau? On the basis of the above, while Taoshan's initiative is likely useful, and

[120] Author's interviews with county BOH Director (June 2010).

possibly a helpful stepping stone to greater oversight, it is doubtful whether this really fulfils the functions one would hope for from significant structural change to local oversight.

Taoshan's initiative and supra-local policy change

Where now for Taoshan's policy? Locally, there is a clear belief that the Bureau performs a useful function. Members of the Bureau and BOH argue for extending its work to include other components of local health system oversight, transforming this from limited NCMS oversight into a much more powerful entity with an expanded role. They argue that this would make sense: the need for oversight is there; the Bureau exists and could quite easily be extended; much more sense to group oversight functions than to separate them. This may happen, but there are no plans for this as it stands. In contrast, as seen from outside the county, Taoshan's initiative may have missed the boat: for a brief moment, the Oversight Bureau had a modicum of provincial and central limelight, but this appears to have passed.

Taoshan's initiative has received a degree of policy attention outside the county. The clearest indicator of this influence was the county being invited to discuss its Oversight Bureau at a provincial meeting attended by other counties, with the hope that these counties would learn from Taoshan's initiative. Other counties in the province have subsequently established similar oversight mechanisms, though it is impossible to determine the extent to which these are linked to Taoshan's IOB, though BOH and Bureau staff clearly believe that they are. The county has also received a degree of central government attention. The BOH was invited to submit a report on their initiative to the MOH in preparation for a national meeting at which they were to discuss their work; however, after a delay, the county was informed that the meeting would not be convened. Reasons are unclear, but locally it is speculated that Taoshan's timing was unfortunate: that this initiative, focused on strengthening NCMS oversight and management, had been overtaken by a shift in central attention away from innovations in oversight towards provider payment reforms in 2009-2010. Both of these events should be understood as clear signals of approbation for Taoshan's reform.

Overall, systemic judgements of Taoshan's reform have been positive, as shown here, but it seems that this reform missed a window of opportunity for being promoted nationally as a model. Interviews with province-level NCMS managers reveal that provincial thinking on development of the NCMS has moved on: Taoshan's oversight model is seen as a useful, but inadequate, solution to NCMS oversight. At the time of interview (June 2010), the provincial BOH was working on developing NCMS payment reforms, seen as more fundamental to the development of the NCMS than Taoshan's initiative.

The need for such solutions remains, however. A very recent MOH document on strengthening NCMS fund management calls for the establishment of just this kind of agency:

> There is a need to build effective mechanisms to separate management and operation [of the NCMS]: each province (district, city) must ...ensure the separation of NCMS management and operation at the township level and THC services. [Counties should] implement a system under which county [NCMS] operating agencies second operational and audit staff to the township level [and] ensure the independence of NCMS operating personnel. Currently, it is hard to separate NCMS management and operation from service provision at the township level, [so] county-level NCMS operating agencies should strengthen work on auditing of reimbursement (Weinongweifa 2011).

This is a very precise description of the kind of solution to oversight problems developed in Taoshan. While the need for a model such as this remains evident, it is impossible to say whether Taoshan's reform will be picked up again and boosted nationally, and it is doubtful whether this initiative could have significant supra-local impact without some sort of higher-level intervention and propagation.

There is a degree of ambiguity in such a clear central exhortation, late in the development of the scheme, for development of management structures basically set out in Taoshan (and elsewhere) several years previously. Not only does this seem to be an admission that this remains a deficit in management of the scheme but more generally it points to the lack of systemic recycling of useful local practices, both Taoshan's management model and equivalent or similar practices developed elsewhere. Saying this, two points must be borne in mind: first, Taoshan-type management practices have achieved at least a degree of sub-national coverage, despite a seemingly low level of specific central attention to this, and part of the reason for this is surely sub-national propagation efforts such as the provincial meeting discussed above. Second, detailed central direction on separation of management and oversight and the promotion of Taoshan-type agencies, late though it is, is likely more important than actual propagation or replication of Taoshan's model *per se*. In other words, while Taoshan and other sub-national models are not being specifically 'rolled out', the underlying idea is clearly articulated as a principle of scheme development.

Conclusion

This chapter has shown one county's development of a local oversight mechanism within the NCMS and how this came to have a degree of supra-local impact. This initiative responds to local concerns, but also highlights deficiencies in scheme management more generally which largely stem from specific national debates on the 'peasant burden' pre-dating the NCMS which were subsequently carried through into scheme design. Taoshan's Oversight Bureau results clearly and directly from this conjuncture and, while the county has experienced some kind of crisis in its recent implementation of the NCMS, the overall issue addressed in Taoshan is structural, not purely local and contingent, and this is clearly visible in the prevalence of risk framings in other counties, and the existence of similar initiatives in other places.

Taoshan's reform should be seen as a product of China's 'distinctive policy process'. 'Piloting' of reforms encompasses a range of different types of activity, from the directly managed and clearly experimental, to sub-national 'tinkering'. Attachment to a discourse of 'piloting' responds to two main beliefs 'in and around government': first, that local starting points and developmental trajectories are necessarily heterogeneous, requiring substantial local discretion within central policy mandates; second, that useful experience gained during reform can illuminate policy development – in the case of the NCMS, that sub-national lessons can feed into national policy. Little attention has been paid to the sub-national dynamics associated with this kind of policy development process. We see these clearly in Taoshan's reform.

While concrete impact of Taoshan's reform has been limited, our reading of what Taoshan has done should focus more on provincial propagation and local spin-offs, both of which point to the local importance of this kind of reform in policy development, and to the way in which this reform highlights a long-standing problem within the scheme. Specific MOH attention to exactly this kind of reform, late though this was in coming, is welcome for confirming a clear direction in development of the NCMS and the usefulness of solutions such as Taoshan's. The importance of this process of tinkering lies in allowing decentred problem solving at the coal face in China's reforms and the possibility of systemic recycling of these practices and models.

Bibliography

Bloom, G., Han L. & Li, X. 2000. How health workers earn a living in China. IDS Working Paper no. 108. Institute of Development Studies, Brighton, UK. Available at: http://www.ids.ac.uk/publication/how-health-workers-earn-a-living-in-china.

Bloom, G. & Jing, F. 2003. China's rural health system in a changing institutional context. IDS Working Paper no. 194, Institute of Development Studies, Brighton, UK. Available at: http://www.ids.ac.uk/idspublication/china-s-rural-health-system-in-a-changing-institutional-context.

Brown, P.H., De Brauw, A. & Du, Y. 2009. Understanding variation in the design of China's New Co-operative Medical System. *The China Quarterly* 198: 304-329.

Chen, F. 2010. Xinxing Nongcun Hezuo Yiliao jiandu zhidu cunzai de wenti ji qi duice tantao [Problems in oversight of the New Cooperative Medical Scheme and investigation of countermeasures]. *Zhonggong Fujian Sheng Wei Dangxiao Xuebao* [*Journal of the Party School of Fujian Province*] 2.

Christiansen, F. & Zhang, H.X. 2009. The political economy of rural development in China: Reflections on current rural policy. Duisburg Working Papers on East Asian Studies. Universität Duisburg-Essen, Duisberg, Germany. Available at:
http://duepublico.uni-duisburg-essen.de/servlets/DocumentServlet?id=22717.

Eggleston, K., Li, L., Meng, Q., Lindelow, M. & Wagstaff, A. 2006. Health Service Delivery in China: A Literature Review. World Bank, Washington D.C. Available at: http://documents.worldbank.org/curated/en/2006/08/6962134/health-service-delivery-china-literature-review.

Gao, G. 2008. Gonggu he fazhan juyou Zhongguo tese de Xinxing Nongcun Hezuo Yiliao baozhang zhidu [Consolidate and develop the New Cooperative Medical Scheme with Chinese characteristics]. *Yiliao weisheng lüpishu: Zhongguo yiliao weisheng fazhan baogao 4* [*Green book of health care: Annual report on China's healthcare, no. 4*]. Du, L. (ed.). Shehui Kexue Wenxian Chubanshe [Social Sciences Academic Press], Beijing, China.

Gao, G. & Han, Y. 2007. Wo guo Xinxing Nongcun Hezuo Yiliao zhidu de fazhan [Development of China's New Cooperative Medical Scheme]. *Yiliao weisheng lüpishu: Zhongguo yiliao weisheng fazhan baogao 3* [*Green book of health care: Annual report on China's healthcare, no. 3*]. Du, L. (ed.). Shehui Kexue Wenxian Chubanshe [Social Sciences Academic Press], Beijing, China.

Gao, G., Zhao, L. & Wei, W. 2006. Xinxing Nongcun Hezuo Yiliao zhidu de yunxing chengben he guanli chengben tantao [Investigation into operation and management costs of the New Cooperative Medical Scheme]. *Zhongguo Weisheng Jingji* [*China Health Economics*], 276: 24-26.

Gao, G. & Zhu, J. 2009. Wo Guo Xinxing Nongcun Hezuo Yiliao jijin guanli guocheng zhong cunzai de wenti yu yingdui celüe [Problems in China's New Cooperative Medical Scheme fund management and policy countermeasures]. *Zhongguo Weisheng Jingji* [*China Health Economics*], 28(1): 52-55.

Guobanfa 2003. No. 3. *Guowuyuan Bangongting zhuanfa Weisheng Bu deng bumen guanyu jianli Xinxing Nongcun Hezuo Yiliao zhidu yijian de tongzhi* [Notice of the Office of the State Council on the forwarding of the Opinions of the Ministry of Health and other ministries regarding the building of the New Cooperative Medical Scheme], 10/01/2003.

Heilmann, S. 2008a. Experimentation under hierarchy: Policy experiments in the reorganization of China's state sector, 1978-2008. CID Working Paper no. 172.

Centre for International Development, Harvard University, Cambridge, MA, USA. Available at: http://www.hks.harvard.edu/var/ezp_site/storage/fckeditor/file/pdfs/centers-programs/centers/cid/publications/faculty/wp/172.pdf.

Heilmann, S. 2008b. From local experiments to national policy: The origins of China's distinctive policy process. *The China Journal* 59: 1-30.

Heilmann, S. & Perry, E.J. 2011. Embracing uncertainty: Guerrilla policy style and adaptive governance in China. *Mao's invisible hand: The political foundations of adaptive governance in China*. Heilmann, S. & Perry, E.J. (eds.). Harvard University Press, Cambridge, MA, USA / London, UK.

Herd, R., Hu, Y.-W. & Koen, V. 2010. Improving China's health care system. Paris, Organization for Economic Cooperation and Development. Economics Department Working Papers no. 751. Available at: http://search.oecd.org/officialdocuments/displaydocumentpdf/?doclanguage=en&cote=eco/wkp(2010)7.

Hu, S., Tang, S., Liu, Y., Zhao, Y., Escobar, M.-L. & De Ferranti, D.. 2008. Reform of how health care is paid for in China: Challenges and opportunities. *The Lancet* 372: 1846-53.

Jackson, S., Sleigh, A.C., Li, P. & Liu, X.-L. 2005. Health finance in rural Henan: Low premium insurance compared to the out-of-pocket system. *China Quarterly* 181: 137-157.

Kingdon, J.W. 1984. *Agendas, alternatives, and public policies*. Little Brown and Company, Boston, USA.

Lam, T.-c. 2010. The county system and county governance. *China's local administration: Traditions and change in the sub-national hierarchy*. Chung, J.H. & Lam, T.-c. (eds.). Routledge, London, UK.

Liu, Y., Rao, K., Wu, J. & Gakidou, E. 2008. China's health system performance. *The Lancet* 372: 1914-1923.

Mao, T.-t. 1977. On the ten major relationships. *Selected Works of Mao Tse-tung, Volume 5*. Foreign Languages Press, Peking, China.

MOH Center for Statistics Information 2007. *Zhongguo Xinxing Nongcun Hezuo Yiliao jinzhan ji qi xiaoguo fenxi: 2005 nian Xinxing Nongcun Hezuo Yiliao shidian diaocha baogao* [*Progress and effectiveness evaluation of the New Cooperative Medical Scheme: 2005 NCMS pilot investigation report*]. Zhongguo Xiehe Yike Daxue Chuban She [Peking Medical University Publishing House], Beijing, China.

NCMS Pilot Evaluation Group 2006. *Fazhan zhong de Zhongguo Xinxing Nongcun Hezuo Yiliao: Xinxing Nongcun Hezuo Yiliao shidian gongzuo pinggu baogao* [*The New Cooperative Medical Scheme under development: New Cooperative Medical Scheme pilot work report*]. Renmin Weisheng Chubanshe [People's Medical Publishing House], Beijing, China.

Nongweisi 2008. *Weisheng Bu Bangongting guanyu queding Xinxing Nongcun Hezuo Yiliao you guan shidian zhongdian lianxi diqu de tongzhi* [Notice of the General Office of the Ministry of Health confirming main linking areas for New Cooperative Medical Scheme pilots], Rural Health Office of the Ministry of Health, People's Republic of China, 27/07/2008.

People's Daily Online, 2007. Guowuyuan Fuzongli Wu Yi: Quanmian tuijin Xinxing Nongcun Hezuo Yiliao fazhan [Vice President of the State Council, Wu Yi: Comprehensively promote the development of the NCMS], 16/03/2007. Available at: http://theory.people.com.cn/GB/49169/49171/5479218.html.

Schurmann, H.F. 1966. *Ideology and organization in Communist China*. University of California Press, Berkeley, CA, USA.

Taoshan County Bureau of Health, 2008. Implementation plan on the establishment of Taoshan's Inspection Bureau. Unpublished document.

Taoshan County Bureau of Health. 2009a. New Cooperative Medical Scheme work handbook. Unpublished document.

Taoshan County Bureau of Health. 2009b. Regulations on Taoshan County's New Cooperative Medical Scheme inspection work. Unpublished document.

Taoshan County Bureau of Health, 2010. Perfect mechanisms, manage scientifically, promote the effective and stable operation of the New Cooperative Medical Scheme). Unpublished document.

Van Doorslaer, E., O'Donnell, O., Rannan-Eliya, R.P., Somanathan, A., Adhikari, S.R., Garg, C.C., Harbianto, D., Herrin, A.N., Huq, M.N., Ibragimova, S., Karan, A., Ng, C.W., Pande, B.R., Racelis, R., Tao, S., Tin, K., Tisayaticom, K., Trisnantoro, L., Vasavid, C. & Zhao, Y. 2006. Effect of payments for health care on poverty estimates in 11 countries in Asia: An analysis of household survey data. *The Lancet* 368: 1357-1364.

Wagstaff, A. & Lindelow, M. 2008. Health reform in rural China: Challenges and options. In: Lou, J. & Wang, S. (eds.). *Public finance in China: Reform and growth for a harmonious society*. The World Bank, Washington, DC, USA.

Wagstaff, A., Lindelow, M., Gao, J., Xu, L. & Qian, J. 2009a. Extending health insurance to the rural population: An impact evaluation of China's New Cooperative Medical Scheme. *Journal of Health Economics* 28:1-19.

Wagstaff, A., Lindelow, M., Wang, S. & Zhang, S. 2009b. *Reforming China's rural health system*. The World Bank, Washington DC, USA.

Wagstaff, A., Yip, W., Lindelow, M. & Hsiao, W.C. 2009c. China's health system and its reform: A review of recent studies. *Health Economics* 18: S7-S23.

Weinongweifa 2008 no. 17. *Guanyu zuo hao 2008 nian Xinxing Nongcun Hezuo Yiliao gongzuo de tongzhi* [On doing New Cooperative Medical Scheme work well in 2008], Ministry of Health, People's Republic of China, 30/03/2008.

Weinongweifa 2011 no. 52. *Guanyu jin yi bu jia qiang Xinxing Nongcun Hezuo Yiliao jijin guanli de yijian* [Opinions on strengthening management of New Cooperative Medical Scheme funds], Ministry of Health, People's Republic of China, 25/05/2011.

Wu, Z. 2009. Lun Xinxing Nongcun Hezuo Yiliao jiandu jizhi chuangxin [Discussion of innovating mechanisms for oversight of the New Cooperative Medical Scheme]. *Zhonggong Fujian Sheng Wei Dangxiao Xuebao [Journal of the Party School of Fujian Province]* 12: 37-43.

Yang, T. 2004. Zhongguo Shehui zhengce xingjin zhong de ji ge jiben wenti: Yi Xinxing Hezuo Yiliao zhengce wei li [Basic problems in China's social policy: The example of the New Cooperative Medical Scheme]. Social Policy Research Centre, Chinese Academy of Social Sciences, Beijing, China. Available at: http://www.chinasocialpolicy.org/Paper_Show.asp?Paper_ID=166.

Yip, W. & Hsiao, W. 2008. Market watch: The Chinese health system at a crossroads. *Health Affairs* 27(2): 460-469.

Yip, W. & Hsiao, W. 2009. Non-evidence based policy: How effective is China's new cooperative medical scheme in reducing medical impoverishment? *Social Science & Medicine* 68: 201-209.

Zhang, L., Liu, Y., Feng, Z. & Song, T. 2007. *Chongqing Shi Qianjiang Qu nongcun weisheng zhengce ji guanli gaige: Anli yanjiu [A case study of rural health policy and management reform in Qianjiang District, Chongqing]*. Zhongguo Caizheng Jingji Chubanshe [China Financial and Economic Publishing House], Beijing, China.

Zhang, L., Yi, H., Luo, R., Liu, C., Shi, Y., Rozelle, S., Babiarz Singer, K., Miller, G. & Atlas, S. 2011. Caring for one billion: Assessing the new social protection programmes in rural China. Brighton, Centre for Social Protection, Institute of Development Studies. CSP Research Report no. 7. Available at:
http://www.socialprotectionasia.org/Conf-prgram-pdf/7-SPA-Final-Paper-No-07.pdf.

Zheng, Y. 2007. *De facto federalism in China: Reforms and dynamics of central-local relations.* World Scientific, Singapore.

Zhongfa 2002 no. 13. *Zhonggong Zhongyang, Guowuyuan guanyu jin yi bu jiaqiang nongcun weisheng gongzuo de jueding* [Decision of the Party Centre and the State Council on continuing to increase rural health work], Central Committee of the Communist Party of China, People's Republic of China, 19/10/2002.

Zhonghua Renmin Gongheguo Shenjishu Bangongting 2011 no. 7. *Zhonghua Renmin Gongheguo Shenjishu shenji jieguo gonggao* [Public announcement of audit results of the National Audit Office of the People's Republic of China], General Office of the National Audit Office of the People's Republic of China, 16/02/2011.

Chapter 9: Horizon of riches

Development and difference in Xinjiang, China

Adam Tyson

Since the formation of the People's Republic of China (PRC) in 1949 there have been three waves of political and economic transition that broadly account for the country's rapid economic rise. The first took place following the death of Mao Zedong (1976), the second was triggered by the Tiananmen Square crisis (1989), and the third is linked to China's formal entry into the World Trade Organisation (2001). In the post-Mao reform era, and from a relatively low baseline, China registered annual economic growth rates of approximately 10% and lifted some 600 million people out of poverty. Capturing this sense of irresistible growth, 'Shenzhen speed' has entered the strata of Chinese economic language. Shenzhen was one of the first Special Economic Zones (SEZs) serving as catalysts for China's economic growth and giving rise to a new urban class of Han millionaires.[121] Set against conventional theories of international political economy, Whyte (2009) exposes the paradoxical nature of this modern economic miracle. Zhao (2010) and Jiang (2011), among others, are advancing new research agendas related to the 'Beijing Consensus' in order to establish whether China offers a plausible, alternative economic model.

China's future economic trajectory largely depends on the performance of the so-called fifth generation leaders. The 18th National Congress of the Chinese Communist Party (CCP) was held in November 2012. President Xi Jinping, Premier Li Keqiang, and the new Central Politburo Standing Committee have signalled their general intentions for the duration of their tenure (2012-2022). They are bound by national economic performance targets set out in the 12th Five Year Plan (2011-2015) and synchronised with the ongoing Western Development Project (WDP). Established in 1999, the WDP serves to narrow China's east-west developmental gap and mitigate against regional distribution disparities. For instance, some prefectures in the Xinjiang Uyghur Autonomous Region (XUAR) on China's conflict prone and resource rich western frontier – are believed to be 'a century behind Guangdong' in developmental terms (Sautman 2012: 25). I accept that Muslim minorities (especially Uyghurs) in Xinjiang stand to benefit from Beijing's WDP projects, subsidies and affirmative action policies, but argue in this chapter that minority groups also view

[121] Han Chinese account for approximately 90%of the population and are the largest of China's 56 officially recognised ethnic groups. The best known ethnic minority groups in Xinjiang are the Uyghur, Kazakh, and Hui. The current population of China has surpassed 1.3 billion.

C. Brautaset and C.M. Dent (eds.) **THE GREAT DIVERSITY**
DOI 10.3920/978-90-8686-786-8_9, © Wageningen Academic Publishers 2014

development differently and do not fully subscribe to or endorse the government's heady embrace of industrialisation, urbanisation, and intensive resource extraction.

There is increasing scope in contemporary China to challenge mainstream development agendas, fight corruption, agitate for environmental protectionism, and advocate for labour rights. Less progress has been made, however, when it comes to the management of ethno-religious minority relations. Minorities in Xinjiang have long been embroiled in political struggles surrounding identity, rights and recognition. International linkages have been formed with diasporas and advocacy groups in European and North American cities (particularly the World Uyghur Congress in Berlin), and solidarity networks based primarily on Islam exist with neighbouring Central Asian countries. As this chapter demonstrates, significant cultural incongruities remain when it comes to definitions of development, progress, and individual betterment in Xinjiang's diverse cities and prefectures.

Beijing has experimented with different policy options to appease and control minorities, although given the salience of security threats the government tends to settle for a coercive approach to the Xinjiang quandary.[122] Rebiya Kadeer, perhaps the best known Uyghur activist, is considered by Beijing to be a dangerous provocateur and terrorist mastermind. In a highly emotive *Wall Street Journal* editorial marking the third anniversary of the 5 July 2009 mass riots in Urumqi (the capital of Xinjiang), appeals were made to 'curb China's brutal reign' (Kadeer 2012: A13). Beijing's securitisation strategy in Xinjiang should be viewed with ambivalence as it serves to contain separatism and enclose minorities as well as to open the region to investment and development opportunities.

In order to shed further light on the CCP's desire to develop their way out of conflict in Xinjiang while simultaneously civilising and modernising minorities such as the Uyghur, this chapter adopts the 'Chinese eyes' approached favoured by Lagerkvist (2009). In his study of Sino-African relations, Lagerkvist (2009: 121) locates and analyses Chinese academic perspectives which are 'largely hidden from the non-Chinese-speaking world'. For the purposes of this chapter, this is done by researching Tsinghua Tongfang's China National Knowledge Infrastructure (CNKI) database, as well as Ori Probe's China-Asia On Demand (CAOD) document delivery service.[123] As revealing as this is, there are several caveats worth mentioning. Aside from the tiered system used to rank academic journals in China, it is difficult to

[122] Xinjiang is a restless region marked by insurgencies, separatist movements, and everyday forms of resistance. Riots on 5 July 2009 (known locally as the 7.5 riots) captured international attention, with estimates of 184 fatalities and thousands of injuries (Clarke 2011).

[123] These databases contain thousands of digitised Chinese-language journals and academic materials.

gauge the influence of individual Chinese scholars, and the extent to which their findings and arguments align with government policy is not always clear. In addition, there are a host of other hidden voices commanding much greater political power and industrial influence in the XUAR, for example senior members of the Xinjiang Production and Construction Corps or regional Party secretaries, although academic viewpoints do offer important lenses into the complex ethno-national and developmental dilemmas on China's northwest frontier.

Chinese social scientists tend to use quantitative methodologies to explain human behaviour and reach generalised conclusions that are expected to have policy relevance and social impact. In China, survey data published by supposedly dispassionate academics is held to be more reliable and less overtly politicised than most media, non-government and activist sources. Nevertheless, Rosen (2003) identifies a range of methodological limitations including the snapshot nature of survey findings (lack of time-series data), the distortion factor, narrow sampling techniques, and the prevalence of self-censorship. The 24 Chinese language journal articles selected for this chapter will be supplemented by personal correspondence with postgraduate students and scholars from universities in China.

This chapter begins with an overview of Xinjiang and the gradual incorporation of this frontier region into what is now the PRC. I will then examine the politics of ethnic difference and cultural incongruities in Xinjiang, revealing problematic aspects of Beijing's horizon of riches strategy. The third section builds on Becquelin's (2004: 361) analysis of theatrically staged development, where a 'horizon of untold riches' is said to await all those in minority, frontier regions willing to subscribe to Beijing's developmental goals and the broader ethos of efficiency and economic productivity. Finally, the fourth section is concerned with the developmental role played by the Xinjiang Production and Construction Corps (XPCC), a powerful agro-military state organ composed of 14 divisions strategically spread across Xinjiang's cities, prefectures and counties.

New frontier

Some 21 million people belonging to as many as 46 different ethnic groups reside in Xinjiang, a region constituting a sixth of China's total land area and bordering eight countries (Clarke 2011). From the 18th century onwards the Turkic-Muslim minorities of the region have witnessed the encroachments of organised Chinese agro-military forces known first as *tuntian* and more recently as *bingtuan*. The region was renamed Xinjiang [literally 'new frontier'] in 1884 and was the site of ongoing contestation between Peking [Beijing] and Moscow. Policies of mass transmigration ensured that Han Chinese retained a strong foothold in the region and guarded all territories inherited from the Qing Dynasty (1759-1911).[124]

Xinjiang fell under direct Soviet control during 1934-1942 after which the region was liberated, or recaptured depending on one's vantage point, by China. Ever since there have been government concerns related to the sustainability of central subsidies and transfer funds required to stabilise Xinjiang, which constitute a large proportion of expenditure in the region. By the 1980s an increasing north-south developmental gap was another cause for anxiety, with the northern part of Xinjiang and the capital Urumqi dominated by Han settlers and expansionists (Shan & Weng 2010). This imbalance was exacerbated in the 1990s by intensified resource extraction and the arrival of energy giants such as PetroChina and Sinopec. Minority discontent spread from mosque to bazaar, and sporadic demonstrations and riots took place in Aksu, Baren, Ghulja, Kashgar, Urumqi, and Yining, among other cities.

In response to increasing unrest, Xinjiang has been made a focal point of Beijing's Western Development Project since 1999, with the XPCC playing a leading role. With many hundred billions of dollars invested so far, the WDP has gone some way to promoting indigenous innovation and enhancing productivity and efficiency, although there have been signs of excess and hubris. For example the burden of the 4,200 km West-to-East Pipeline project designed to deliver natural gas from Xinjiang to Shanghai has fallen on PetroChina (majority owned by the Chinese government), and researchers have determined that these gas shipments 'cost 45% more than international imports' (Millward 2007: 302). As such, the rationale must have been to stimulate growth in Xinjiang as well as to gain political capital by demonstrating that the WDP is operational, suggesting that regional stability comes at a hefty price.

Xinjiang also features in China's 12th Five Year Plan (2011-2015) with goodwill themes such as *baorongxing zengzhang* [inclusive growth] and *minzu tuanjie* [ethnic integration/unity], along with pragmatic appeals for *suoxiao diqu chabie* [regional economic rebalancing], agricultural innovation, and increasing investment in infrastructure. The Asian Development Bank has been assisting in this process, funding a range of infrastructure projects, integrated urban development, and environmental improvement in Xinjiang.[125] Uyghurs and other minorities have not,

[124] State-directed immigration to increase the number of Han Chinese in Xinjiang, sometimes referred to as the 'trap policy' of Hanisation, brought the figure to approximately 8.2 million in 2007, and this continues to rise (Bovingdon 2010).

[125] See for instance Asian Development Bank Project 45508-001. Earmarked at US$900,000 this project was approved in November 2011 and includes a range of technical support services related to water treatment, ecological rehabilitation, and infrastructure network expansion in the second-tier cities of Kelamayi and Kuitun.

however, fully enjoyed the residual benefits of these development plans, nor have they fully participated in the coal, oil and gas exploration in their prefectures, which fuels China's special economic zones and creates a strategic 'energy junction' between Central Asia, South Asia and East Asia (Clarke 2011: 151).

Members of the Central Politburo parsimoniously claim that development holds the solution to all Xinjiang's problems and that the advanced social forces (meaning productive, dexterous citizens) driving development are the most 'desirable' citizens (Barabantseva 2009: 229). Generous university scholarships have been offered to minorities, producing an upwardly mobile elite with a newfound appreciation for Chinese (Han) civilisation, fluency in Mandarin, and access to ever-important *guanxi* business networks and political connections. Nevertheless, ethnic and religious minorities retain a sense of otherness, and many feel as though they are in the grip of the CCP's highly opaque and prejudiced language and education policy (Schluessel 2007). In peripheral regions such as the Guangxi Zhuang Autonomous Region and the XUAR, minorities often have their own ideas about individual betterment, religious devotion, progress, and the appropriate pace of development.

It is not surprising that efforts to reorient and rebalance China's development strategy, including the WDP, have been met with formidable challenges. These range from government financial constraints, hesitant foreign investors, rampant corruption, poor human resources, logistical barriers, ecological concerns, and separatist movements in Xinjiang, Tibet and elsewhere. Authors such as Tian (2004: 636) invite controversy by adding that the 'unhealthy mentality' of Xinjiang's local population is to blame for the regional developmental deficit and dependency on subsidies. This rather crude view is still prevalent in Xinjiang today. Drawing on ethnographic field research and a wide range of interviews in Urumqi, for instance, Caprioni (2011) reveals a pattern of negative stereotyping with minorities viewed as second-class backward citizens, and ethnic Han Chinese represented as heartless workaholics by minority ethnic groups.

As with all Chinese citizens, Uyghurs, Kazakhs and other minorities are endowed not only with the 'right to development' but also the obligation and duty to contribute economically and to achieve individual betterment through productivity and hard work (Barabantseva 2009: 227). With this in mind, officials in Xinjiang are known to 'counterpoise piety to development', resulting in prohibitive measures and the curbing of religious activities in public schools and workplaces (Sautman 2012: 16). Pious Muslims have constitutionally guaranteed rights, although in practice there are limits to state tolerance, and in Xinjiang officials are concerned that some religious practices have deleterious social effects and are a drain on development.[126] Examples include the daily call to prayer led by local *muezzin*, periodic fasting (especially Ramadan) and sacrificial rituals, and the controversial usage of 'Xinjiang time' (two hours behind Beijing) allowing the non-Han population to align their meals and daybreaks with prayer times rather than standard working schedules or office hours. Young men under the age of 18 are discouraged from spending too much idle time in mosques. Everyday pressures to acculturate are linked to the promise of wealth and opportunity in a rapidly modernising China.

Cultural incongruities

Context

Xinjiang, host to more than 10 million Turkic-Muslims, was deeply affected by the independence movements in neighbouring Central Asia at the end of the Cold War (Guo & Chen 2012). By the mid-1990s separatist groups were agitating for the formation of an independent East Turkestan to replace their status as a dominion or new frontier dominated by Han Chinese. It is important to question the representativeness of separatist groups as the majority in Xinjiang seem to favour non-violent resolution reached through political processes and negotiation, although expressions of ethnic difference are often narrowly read by government authorities and cadres as a threat to national unity.[127] Chinese police have responded by formulating sophisticated mechanisms for dispute resolution, mediation and the quelling of unrest. Guo & Hu (2012) provide a rich account of the new policing mechanisms being tested in Kashgar, western Xinjiang.

Official anxieties towards Xinjiang's many minority groups can also be viewed in relation to an emerging field known as hyperlink network analysis, whereby virtual social bonds and transnational affinities are forged, and new avenues of digital coordination and activism are created (Sullivan & Xie 2009). Using a comparative framework, Xiu-li Wang (2012) examines the threat to all those in Xinjiang who are exposed to the freedoms of unregulated cyberspace, suggesting that the whole region is under ideological attack by 'Western' (particularly US) agents. Enhanced vigilance is therefore deemed essential, and authorities are encouraged to redouble efforts to modernise Xinjiang on the basis of *kexue fazhan guan* ([Scientific Outlook on Development], coined by former president Hu Jintao) and reified versions of minority culture.

---

126 See for instance Article 36 of the 1982 Constitution of the PRC.

127 Personal correspondence with academics from Xinjiang University, November 2012. The anonymity of all respondents is protected.

Given the disparate views of success, achievement, religious values, and the work-life balance found in many of Xinjiang's cities, top-down developmental agenda setting is problematic. A good example is the improbable 'love the great motherland and create a better home' Xinjiang initiative launched by Li Changchun and Zhou Yongkang, both of the Central Politburo, in June 2010 (Guo & Dong 2011: 59). In 2012, by contrast, a RMB300 million (US$48.5 million) teacher encouragement pilot project was announced, targeting some 15,000 teachers with financial difficulties (CPT 2012). Aside from the sensible project title, this resonates (to an extent) with local needs and aligns with the reform agenda of Nur Bekri, who has served as Chairman of the Xinjiang Uyghur Autonomous Region since 2007. Nur Bekri, a Uyghur, is in charge of local governance although real power is held by his Han superior, the CCP Secretary Zhang Chunxian (transferred from Hunan Province in 2010). Facing both active and latent separatist movements, Xinjiang's leaders face formidable challenges when it comes to balancing security imperatives with stringent economic growth targets and the achievement of inter-ethnic reconciliation.

Rethinking Minzu

Article 4 of China's 1982 Constitution holds all *minzu* [nationalities] to be equal under law. Rights are guaranteed, discrimination is punishable, and autonomous regions are inalienably part of the PRC. Provisions for ethnic minorities are further codified in the 1984 Law on Minority Regional Autonomy (Chen 1999). There are, however, only 56 recognised *minzu* across China, suggesting a highly proscriptive and discretionary process of selection informed by regional politics, territorial integrity, energy imperatives, and security considerations. Official recognition is part of CCP strategy to sever ties to local or tribal groupings, building instead 'loyalty to a larger collectivity' under the paternal care of the government (Kaup 2000: 25). Even the Han are said to have been created through historical processes of assimilation and the intermixing of multiple disparate tribes, reminiscent of colonial legibility projects in the Dutch East Indies (Indonesia) and British Malaya. In contrast to the selective process of ethnic classification and enclosure, China is facing an emergent 'politics of unofficial ethnicity' in which hundreds of sub-ethnic groups and *gaoshan* [indigenous] identities are being rediscovered and linked to rights-based struggles (Gladney 2004: 23).

Academics from Xinjiang confirm that alternative views on development and progress held by the region's Muslim minorities are rarely discussed openly by researchers or public officials.[128] Dissenting or contrarian views are deemed

[128] Personal correspondence with academics from Xinjiang University, November 2012.

sensitive by the government, especially when it comes to political, religious, national or ethnic issues. A vast censorial and surveillance apparatus exists in Xinjiang, ensuring that university researchers focus primarily on minority cultural, artistic or educational issues that are largely devoid of politics. Students of contemporary China will recognise what Guo (2008: 350) calls the 'sensitivity gap', which refers to the difference between availability of information and the freedom or willingness to analyse information if and when it has been disclosed.

When it comes to the permissible aspects of *minzu*, Chinese officials keenly showcase Xinjiang *muqam* traditions in all their artistic manifestations, encouraging research into the aesthetic bonds forged through music, song and dance (Sun & Zhao 2012). More covertly, there is parallel research being conducted into the potential political manifestations of cultural revivalism. Questions of ethnic relations are taken very seriously by Chinese authorities, and beyond 'culture as art' there is a recognised need to reconcile social misunderstandings, prejudices, and dominant stereotypes between Han trans-migrants and ethnic minorities.

A select number of Chinese scholars have begun to openly question the appropriateness of the politically-loaded term *minzu*, proposing instead the use of *zuqun* when referring to ethnic groups (Wen 2010). This is not simply a question of pedantic zeal, but rather has implications for group rights and recognition in autonomous settings. Consistent with the *he er butong* [unity in diversity] principle, China is held to be one nation with multiple ethnic groups bound by a common language and committed to the territorial integrity of the PRC. Avoiding the political implications of the *minzu-zuqun* debate as well as the possible implications for minority rights in peripheral regions such as Xinjiang, authors such as Wen (2010) instead marshal their evidence towards the rejection of so-called splitist views that threaten to undermine national unity in China.

Nationalism

Continuing with the 'Chinese eyes' approach, there are examples of theoretical engagement with classic Western studies on nationalism, ethnicity and identity. Not without controversy, Jiang and Chu (2012) selectively cite the likes of Benedict Anderson and Ernest Gellner in support of the notion that the original inhabitants of Xinjiang must embrace (Han Chinese) modernity and norms through re-education and scientific innovation.[129] Similarly, using a rather outdated and misleading

[129] Specifically Anderson's *Imagined Communities* and Gellner's *Nations and Nationalism*, both originally published in 1983, are cited.

Marxist framework, Ma (2012) traces the emergence of minority identities and cultures, grapples with the challenges of pan-Turkism, and gives an account of contemporary struggles related to nation-building and processes of value diffusion. The analysis ultimately falls short, however, as what Ma (2012) superficially terms 'Xinjiang spirit' is premised on highly abstract notions of fairness and solidarity. The only evidence of this unique spirit is the fact that Xinjiang people 'love their land' – mountains, rivers, and agricultural produce (especially Turpan grapes and Hami melons) – and supposedly rejoice in folk songs such as 'Our Xinjiang is a Beautiful Place'.[130]

Xu & Wu (2012) use a political philosophy approach to analyse attitudes, outlooks and frames of reference of university students in Xinjiang, with the ultimate goal, it seems, of enhancing Marxist ideology and, by extension, regime legitimacy. One of the key concerns appears to be cognitive dissonance and the likelihood of social compliance versus resistance based on an elite sample of (potentially dangerous) ethnic minority undergraduate students. Framed by similar concerns, Feng & Liu (2012) analyse 428 national identification surveys compiled by students in Ili Prefecture. Attempts are made to highlight and flag those demographic segments appearing to favour local or regional affiliations, where ethnic identity markers interfere with the creation of a unified national identity. There are three discernible dimensions to the survey: (1) ethnic identification, (2) ethnic cohesiveness, and (3) ethnic manners and behaviours. Each dimension is said to be cross cut by language, custom, identity, and cultural factors. Students in Ily were asked to rank, on a scale of 0 to 4, the following sample questions (not all survey questions were published by the authors):

- I feel most comfortable when I hear my own ethnic language being spoken.
- The annual Spring Festival is shared and celebrated by the nation as a whole.
- I can distinguish my own ethnicity from others.
- I am keen to learn about the positive attributes of the great Chinese culture and nation.

Feng & Liu (2012) reach very broad conclusions, celebrating China's 'unity in diversity' and the acceptance by students of their civic duties under the umbrella of Chinese nationalism. Concerns are also raised, however, as students are said to lack of maturity, which causes confusion about identity and renders them impressionable and easily manipulated. From a different but related angle, Zhou (2008) attempts to demystify Islam, and Sufism in particular, in Xinjiang by deploying what is presented

[130] This song was written by Ma Hanbing in 1950 as part of the nationalist propaganda campaign by the newly empowered CCP.

as Freudian psychoanalysis. Zhou (2008) manages to identify the ways in which Sufism impacts on minority customs, beliefs and behaviours in Xinjiang, although his work is highly descriptive and does little to challenge or critically engage with official views linking religious piety to underdevelopment.

In order to achieve unity in diversity, a goal shared by many of China's East Asian neighbours, Yujun Wang *et al.* (2012) assess the prospects for interethnic harmony and improved relations between Han and local minorities (particularly Kazakhs). This is based on an analysis of social interactions, status, proximity, and the dynamics of intermarriage in Xinjiang, a highly contentious issue that deserves more than the positive spin the authors have chosen to place on their empirical findings. Survey data from Yujun Wang *et al.* (2012) shows low levels of resistance to the notion of inter-marriage between Han Chinese and Muslim minorities, which indicates that mutual respect and harmonious relations – both conveniently government slogans – are now a reality in Xinjiang. This, we are told, bodes well for the future prosperity of the XUAR.

Conducting a more thorough and critical study concerning Han-Uyghur relations, Sheng (2012) reveals the fragility of ethnic relations in Xinjiang by using mixed methodologies (including field interviews) to gauge local opinions and attitudes towards inter-marriage, social integration, segregation in schools, and the value of religious beliefs. Sheng (2012) concludes with an appeal for new thinking and fresh ideas about the Xinjiang dilemma, and it is no coincidence that his work is published in the *Northwestern Journal of Ethnology*, the Chinese journal of choice for many foreign researchers investigating sensitive political issues.[131] It is clear from the literature surveyed so far that cultural incongruities are set to continue to complicate Beijing's horizon of riches strategy.

Paramilitaries and development

When pressed on issues of human rights, minority empowerment and developmental deficits, Chinese government officials and delegates usually insist that Xinjiang is a 'non issue' and refuse to engage in serious discussions about the politics of their western frontier.[132] This is despite the fact that the July 2009 riots in the capital Urumqi prompted the central government to take unprecedented measures to resolve the Xinjiang quandary. Beijing is said to have deployed 500 officials from

[131] For instance, researchers from the Max Planck Institute, in cooperation with the Chinese Academy of Social Sciences, chose to publish a selection of articles from their 'Religions and Chinese Contemporary Society' forum, held in Beijing from 7-9 December 2011, in this journal.

[132] Interview with a professor from the Shanghai Institute of Social Sciences, May 2012.

64 government departments to investigate sources of ethnic tensions and make recommendations for the appointment of new leaders at regional and prefecture levels Yong (2010). The exact findings of the mission were not publicly disseminated, although academics working in Xinjiang note that the government still fails to accept alternative readings of development based on local minority interpretations and aspirations.[133]

Nur Bekri, the current XUAR Chairman and CCP confidant, made the Uyghur question a top priority during the 12th meeting of the 11th Standing Committee of the Xinjiang Regional People's Congress held just weeks after the July 2009 riots. In order to safeguard ethnic unity the government announced plans to promote civic education and national unity, to tackle poverty, and to provide psychological assistance to victims of the riots (Li 2009). In many ways this mirrors government policy in the aftermath of the 1989 Tiananmen Square crisis, when Deng Xiaoping promised widespread education reform and focused on indoctrination and renewed patriotism. It is difficult to find anyone in China who would take comfort from the notion of a government sponsored psychological assistance programme. A resolution was passed at the end of the Xinjiang Regional People's Congress on 24 July 2009 promising to crack down on violent crimes and terrorist activities. Zhang & Song (2010) revealed that 9,000 cadres were deployed in Urumqi alone, tasked with a range of community policing and outreach activities that, despite the positive public relations spin, constitute surveillance by other means.

In tandem with Party Secretary Zhang Chunxian's securitisation agenda (continuing the 'strike hard' policy against terrorists launched by his predecessor), Nur Bekri acknowledges the need to balance security imperatives with economic growth, and agribusiness with ecological and cultural protectionism (Wei 2011). Specifically, there are concerns about underinvestment in cultural infrastructures (education, arts) in Uyghur-majority prefectures, and efforts are being made to propel Xinjiang to the frontline of China's green development movement. Xinjiang is said to enjoy the laudatory title 'home of fruits and melons', and many Chinese citizens living outside of the region do indeed associate Xinjiang with such palatable delights (Han 2012). In line with green development targets, academics from China's many finance and economics universities are working to forge new public-private networks in order to rebrand Xinjiang, focusing on green and organic foods and implementing new cold chain logistics to ensure that the region's seasonal produce reaches far-flung markets (Han 2012).

[133] Personal correspondence with a postgraduate student from Xinjiang University, November 2012.

In more generic terms, officials tasked with the pacification and enclosure of Xinjiang tend to highlight their commitment to 'leapfrog' development (Weihua & Wei 2011). Penggang Wang *et al.* (2012) argue for the establishment of participatory decision making mechanisms, urging Party committee members at all levels not only to solicit but also to listen to the views of a wide range of stakeholders concerned with regional rebalancing, development and economic planning. One can read this as an appeal to pay greater attention to the views and concerns of minority groups in Xinjiang, although the authors do not make this explicit. With the aim of addressing uneven regional development, Ji *et al.* (2012) deploy a partnership for development approach, examining the complexity of multilayered governance structures in Xinjiang and the policy blockages that occur as a result of competing interests. The obvious implications for power and politics in decentralised settings are not examined, however, as this is deemed overly sensitive.

Following the WDP announced by President Jiang Zemin and Premier Zhu Rongji in 1999, leapfrog development is being spearheaded by the controversial Han-dominated XPCC, first established in 1954 and now boasting some three million members. Most Chinese-language studies concerning the XPCC focus on government approved 'safe' topics such as economic productivity, energy security, population control, environmental footprints, and agricultural innovation. Beginning with the fact that the XPCC is the largest paramilitary agribusiness in China, Wei *et al.* (2010) examine the core functions of the Corps, which range from land reclamation to border defence and paramilitary activities. XPCC leaders are much more than just drivers of industrialisation, patent holders and brand managers, and scientific innovators (particularly bio-engineering). The Corps holds the key to the pacification of Xinjiang and ensures that the region remains a strategic gateway to the west via Central Asia.

In a less ambitious but related study, Lu (2012) examines the role played by the XPCC in land management and warns of the ecological impact of continued urbanisation and industrial expansion. Academics such as Lu (2012) rarely examine minority rights, nor do they openly question the legality of land reclamation or the politics of compensation. Focusing on the talent and professionalism of the XPCC's human resources, Tang and Hu (2012) make a clear case for improved training standards, education and managerial approaches without delving into the complexities of affirmative action, preferential hiring policies, or allegations of ethnic discrimination. These studies are at once sophisticated and uncontroversial, making them broadly representative of Chinese research concerning the XPCC.

As a point of departure from the mainstream studies cited above, Zhiping Wang (2012) focuses on literary representations of women in the XPCC, using popular novels to explain the historic positions and future prospects of women within this vast and complex organisational network. Evidence comes from such works as Deng Pu's 1963 novel *Army Daughters*, which offers a realistic portrait of women growing up in XPCC compounds, including experiences of objectification in highly militarised settings. Over time we are told that conditions improved and by the 1980s women were for the most part emancipated, regaining their dignity. Han Tianhang's 2008 book *Mother and Child* was also examined by Zhiping Wang (2012) in order to shed light on the sacrifices made by XPCC mothers, along with their selfless contributions to the country (supporting husbands and raising children). Whereas the achievements of women are celebrated in popular novels and propaganda films, Zhiping Wang (2012) concludes that nothing can ever fully compensate the countless female Han trans-migrants in XPCC complexes for their sacrifices and losses.

With a broader set of parameters and a very different methodological approach, Wu *et al.* (2012) examine the cultural differences between Han trans-migrants, XPCC employees and local indigenous farmers, although their analysis is limited to the use of survey data (134 households) using Statistical Package for Social Sciences software to determine levels of productivity and contributions to development, framed as civic duty. Wu *et al.* (2012) are either unable or unwilling to engage with pressing political questions such as the extent to which Xinjiang's Muslim minorities reject government developmental policies and view economic growth and productivity targets as an imposition, or what, aside from coercive measures, can be done to reconcile competing viewpoints on development and progress.

In a study bordering on the subversive, Zhang and Li (2012) set out to challenge the legitimacy of the XPCC, although the conclusions drawn are actually quite conservative and the authors dare not extend their critique to government policy or leadership. In a highly qualified manner, the authors acknowledge that the XPCC might have lost its original mandate in Xinjiang given the fact that China is no longer in a state of war.[134] Readers are reminded however that religious extremists continue to cause unrest and pose a real threat to the stability of Xinjiang. Zhang and Li (2012) argue that XPCC operations are not entirely compatible with China's increasingly open and competitive economic system, as their mode of productivity is outdated (still largely planned and subsidised) and unresponsive to market trends. Problems of governance are also raised, as local authorities in Xinjiang are said to

[134] The civil war between the Communist CCP and Nationalist KMT came to an end in 1949. In the early 1950s one of the primary instructions for XPCC workers was to keep 'one hand on the gun, one hand on the pickaxe' (Millward 2007: 252).

complain that the XPCC is a competitor putting a strain on local development and causing conflicts of interest. These local complaints appear as direct quotes in the text, but they remain unattributed, making Zhang and Li's (2012) evidence difficult to crosscheck or verify. While implying that the XPCC is riddled with inefficiencies and a culture of corruption, the authors do not make this explicit in their writings. Indeed, it appears as though their analysis is meant to elicit a response from the XPCC is order to bolster the position and viability of the Corps within Xinjiang.

Conclusion

Set against the current feverish pace of economic growth and modernisation, the term *xianqing* refers to a form of leisure celebrated in China as far back as the 7th century, at the height of the Tang Dynasty (Laughlin 2008). The leisure activities of Muslim minorities in contemporary Xinjiang, by contrast, are often associated with laziness and cultural backwardness, and as such are easily denigrated by Communist Party cadres, regional officials, and XPCC leaders alike. Tensions between trans-migrants and local minority groups remain high, although influential CCP university committees and journal editorial boards ensure that most scholars working in China are unable or unwilling to critically examine ethnic relations. Most research topics are pre-assigned and government funding is linked to fashionable political themes and the latest concepts being developed by CCP leaders. As one colleague from China put it, the most controversial topics in Xinjiang are neglected, and academic freedom remains a distant dream.

The blacklisting of 13 US-based scholars (known as the 'Xinjiang 13') as a result of the 2004 edited volume *Xinjiang: China's Muslim Borderland* serves as a reminder of the risks taken by foreign researchers interested in China's volatile western frontier. Among those barred from China are Dru Gladney and Gardner Bovingdon. Apolitical research topics such as the musicology of the *muqam* are permissible, along with subjects related to non-threatening aspects of minority culture, cuisine and arts. Research articles from the CNKI and CAOD Chinese-language databases reveal that Chinese social scientists tend to rely on quantitative methods to study Han-Uyghur relations and minority behaviours. Caprioni (2011) found significant distortions, however, as local government officials do not seem to hesitate to intervene in the research process, modifying questionnaires or withholding permissions for qualitative fieldwork.

In order to see Xinjiang through 'Chinese eyes' this chapter surveyed 24 Chinese-language journal articles. Mindful of Caprioni's (2011) warnings as well as the gap between empirical observations and the freedom to openly analyse such findings as identified by Guo 2008), my chapter reveals the following patterns. Xinjiang's supposedly natural place (as a new frontier within the borders established during the Qing Dynasty (1759-1911) is taken as given and is actively promoted in a number of Chinese studies. Within these borderlands ethnic difference is acknowledged and treated as a developmental challenge to be overcome by sound policy, good governance, and the promise of a horizon of riches. The WDP is perhaps the best known manifestation of this strategy, although China's minorities do not fully subscribe to the goals and benchmarks of the WDP. This might explain why so many Chinese scholars have been commissioned to study minority identities, student predispositions, and the factors linked to loyalty.

Most studies concerned with nationalism, identity and political psychology in Xinjiang deploy outdated frameworks, and references made to non-Chinese theorists such as Anderson and Gellner are often misleading. On the surface efforts by Chinese scholars to replace *minzu* with *zuqun* when referring to ethnicity should be well received, although it turns out that the objective is not to establish deeper understandings of Xinjiang's ethnic minorities but rather to establish more sophisticated methods of political control and surveillance. Perhaps the most controversial aspect of the research on Xinjiang is the supposedly negative relationship between religiosity and productivity. Such views are increasingly difficult to reconcile so long as any serious debate about the role of political Islam is avoided. In the absence of open and critical debate it is likely that local antagonisms will continue to increase. Research into cold chain logistics, the business and governance structures of the XPCC, and responsible community policing initiatives, though highly advanced, will do little to alleviate ethnic tensions in Xinjiang so long as Chinese scholars are forced into self-censorship and compliance with pre-approved 'safe' topics of enquiry.

Bibliography

Barabantseva, E.V. 2009. Development as localization: ethnic minorities in China's official discourse on the Western Development Project. *Critical Asian Studies* 41(2): 225-254.

Becquelin, N. 2004. Staged development in Xinjiang. *The China Quarterly* 184(June): 358-378.

Bovingdon, G. 2010. *The Uyghurs: strangers in their own land.* New York: Columbia University Press, New York, NY, USA.

Caprioni, E. 2011. Daily encounters between Hans and Uyghurs in Xinjiang: Sinicization, integration or segregation? *Pacific Affairs* 84(2): 267-287.

Chen, J. 1999. *Chinese law: towards an understanding of Chinese law, its nature and development.* Kluwer Law International, The Hague, the Netherlands.

Clarke, M.E. 2011. *Xinjiang and China's rise in Central Asia: a history.* Routledge, Abingdon, Oxford, UK.

CPT, 2012. Communiqué. *China Population Today* 1: 35.

Feng, H. & Liu, Y. 2012. 新疆伊犁州小学儿童民族认同调查研究 [An investigation of the national identification of pupils in Xinjiang's Ili Prefecture]. *Journal of Xinjiang University Philosophy, Humanities & Social Science)* 40(2): 86-91.

Gladney, D.C. 2004. *Dislocating China: Muslims, minorities, and other subaltern objects.* University of Chicago Press, Chicago, IL, USA.

Guo, B. & Hu, F. 2012. 少数民族地区大调解机制运行的现实困境及思考: 以新疆喀什地区为研究样本 [Realistic predicaments and considerations of intermediation mechanisms in minority regions: the case of Kashgar in Xinjiang]. *Journal of Xinjiang Police Officers' Academy* 32(1): 19-25.

Guo, T. & Dong, X. 2011. 社会转型中的嬗变与应对: 谈当前新疆宗教工作及宗教学研究 [Changes and countermeasures in periods of social transitions: a discussion on religious affairs and the science of religion in Xinjiang]. *Social Sciences in Xinjiang* 6: 59-63.

Guo, Y. 2008. Corruption in transitional China: an empirical analysis. *The China Quarterly* 194 (June): 349-364.

Guo, Y. & Chen, F. 2012. 新疆民族关系影响因素探究 [Study of the factors influencing national relations in Xinjiang]. *Journal of Yangtze University(Social Sciences)* 35(3): 98-99.

Han, Y. 2012. Strategic research of Xinjiang forestry and fruit industry based on the 'smile curve'. *Asian Agricultural Research* 4(3): 10-14, 18.

Ji, F., Chen, J., Yuan, F. & Chen, W. 2012. 府际关系视角下的跨区域经济合作: 以江苏对口支援新疆伊犁哈萨克自治州州直地区为例 [Cross-region economic cooperation from an intergovernmental relations perspective: a case study of partner assistance from Jiangsu to Ili Kazakh Autonomous Prefecture]. *Arid Land Geography* 35(3): 494-502.

Jiang, G. & Chu, J. 2012. 国族内的 '他者': 战时新疆族群书写与国族建构 ['The other' in the nation: Xinjiang ethnic writings and the construction of the nation during the anti-Japanese war]. *Guizhou Ethnic Studies* 33(3): 107-111.

Jiang, Y. 2011. Rethinking the Beijing Consensus: how China responds to crises. *The Pacific Review* 24(3): 337-356.

Kadeer, R. 2012. Beijing stirs ethnic hatred in Xinjiang. *The Wall Street Journal* July 3, 2012.

Kaup, K.P. 2000. *Creating the Zhuang: ethnic politics in China.* Lynne Rienner Publishers, Boulder, CO, USA/London, UK.

Lagerkvist, J. 2009. Chinese eyes on Africa: authoritarian flexibility versus democratic governance' *Journal of Contemporary African Studies* 27(2): 119-134.

Laughlin, C.A. 2008. *The literature of leisure and Chinese modernity.* University of Hawai'i Press, Honolulu, HI, USA.

Li, L. 2009. Old mosque's new life. *Beijing Review* August 6, 2009.

Lu, X. 2012. 基于生态足迹的区域生态适度人口研究: 以新疆生产建设兵团为例 [Optimum regional population based on the ecological footprint: the case of the Xinjiang Production and Construction Corps]. *Areal Research and Development* 31(4): 157-160.

Ma, F. 2012. 新疆地域文化视阈下的共有精神家园建设 [Constructing a common spiritual homeland from a Xinjiang regional culture perspective]. *Journal of Xinjiang Normal University (Social Sciences)* 33(1): 29-34.

Millward, J.A. 2007. *Eurasian crossroads: a history of Xinjiang*. Columbia University Press, New York, NY, USA.

Rosen, S. 2003. Chinese media and youth: attitudes towards nationalism and internationalism. In: Lee, C. (ed.). *Chinese media, global contexts*. Routledge, Abingdon, Oxford, UK.

Sautman, B. 2012. Paved with good intentions: proposals to curb minority rights and their consequences for China. *Modern China* 38(1): 10-39.

Schluessel, E.T. 2007. 'Bilingual' education and discontent in Xinjiang. *Central Asian Survey* 26(2): 251-277.

Shan, W. & Weng, C. 2010. China's new policy in Xinjiang and its challenges. *East Asian Policy* 2(3): 58-66.

Sheng, J. 2012. 新疆疏勒县民族关系: 调查及比较研究 [Survey and comparative research on ethnic relations in County Shule, Xinjiang]. *Northwestern Journal of Ethnology* 72: 5-13, 28.

Sullivan, J. & Xie, L. 2009. Environmental activism, social networks and the Internet. *China Quarterly* 198 (June): 422-432.

Sun, X. & Zhao, N. 2012. 新疆麦盖提县刀郎麦西来甫的传承与保护 [The inheritance and protection of the Dolan Mashrap in Maigaiti County, Xinjiang Uygur Autonomous Region]. *Journal of Original Ecological National Culture* 4(2): 131-135.

Tang, W. & Hu, J. 2012. 基于区域异质性理论的兵团人力资源: 开发与管理问题研究 [The development and administration of human resources in the Xinjiang Production and Construction Corps based on the theory of regional heterogeneity]. *Journal of Shandong Institute of Commerce and Technology* 12(1): 9-11, 17.

Tian, Q. 2004. China develops its west: motivation, strategy and prospect. *Journal of Contemporary China* 13(41): 611-636.

Wang, P., Chang, Q. & Li, W. 2012. 基于屯垦戍边使命的新疆兵团人口发展战略研究 [Population development strategy of the Xinjiang Production and Construction Corps based on the settlement defence mission]. *Northwest Population Journal* 33(4): 71-77.

Wang, X. 2012. 新疆网络安全领域意识形态建设之思考 [Considerations of ideological construction in the network security of Xinjiang]. *Journal of Xinjiang Police Officers' Academy* 32(1): 26-29.

Wang, Y., Ge, Y. & Li, C. 2012. 和谐社会语境下民族交往探析: 以新疆昭苏县几个哈汉聚居区为例 [Towards ethnic group exchanges in harmonious society: the case of Kazakh and Han compact communities in Zhaosu County, Xinjiang]. *Journal of Xinjiang Normal University (Social Sciences)* 33(2): 60-65.

Wang, Z. 2012. 新疆生产建设兵团 文学叙事中的女性角色 [The role of women in the literary narrative of the Xinjiang Production and Construction Corps]. *Nankai Journal (Philosophy & Social Sciences)* 4: 59-66.

Wei, S. 2011. Xinjiang builds foundation for rapid development. *China Daily* March 3, 2011.

Wei, T., Zhang, J. & Liang, J. 2010. 兵团农垦团场在农业产业化中的作用 [The role of XPCC farms in agricultural industrialisation]. *Journal of Xinjiang University (Philosophy, Humanities & Social Sciences)* 38(6): 17-20.

Weihua, M. & Wei, S. 2011. Three goals to realise leapfrog development. *China Daily* June 1, 2011.

Wen, M. 2010. 政治斗争中的民族话语: 兼谈'族群'与'民族'概念之争 [The terms for nationality in political struggle: the 'zuqun' versus 'minzu' debate]. *Open Times* 6: 53-66.

Whyte, M.K. 2009. Paradoxes of China's economic boom. *Annual Review of Sociology* 35: 371-392.

Wu, N., Hu, X. & Qiao, J. 2012. 不同体制下农户农田生产活动差异分析: 新疆地方村庄与兵团连队的对比 [Farmland production activity differences among rural households under different administrative systems: comparative analysis between local villages in Xinjiang and the Xinjiang Production and Construction Corps]. *Resources Science* 34(8): 1469-1476.

Xu, Y. & Wu, X. 2012. 新疆少数民族大学生民族观形成的归因分析: 基于态度形成理论的视角 [Attribution analysis on the formation of Xinjiang minority college students national outlook: from the perspective of attitude formation theory]. *Journal of Beifang University of Nationalities* 1: 78-82.

Yong, L. 2010. An economic band-aid: Beijing's new approach to Xinjiang. *China Security* 6(2): 13-23.

Zhang, F. & Li, J. 2012. 新疆兵团屯垦戍边合法性及潜在合法性危机探析 [On the potential legitimacy crisis of the Xinjiang Production and Construction Corps]. *Journal of Central South University (Social Science)* 18(2): 101-105.

Zhang, K. & Song, H. 2010. 从'7-5'事件后的维稳看我区反恐防范: 工作机制的转型升级 [The transformation of counterterrorism in Xinjiang: stability and working mechanisms after the '7-5 riots']. *Journal of Xinjiang Police Officers' Academy* 30(3): 11-15.

Zhao, S. 2010. The China model: can it replace the Western model of modernization? *Journal of Contemporary China* 19(65): 419-436.

Zhou, P. 2008. 新疆苏非派的神秘性探析 [On the mystery of Xinjiang Sufism]. *Journal of Guizhou Normal University (Social Science)* 1: 94-97.

Chapter 10: Mapping the ethical terrain of Chinese aquaculture

Exploring sustainable aquaculture trade in a global context

Scott Bremer, Janne Johansen, Simen Øyen, Matthias Kaiser and Arne Sveinson Haugen

The globalisation of food markets has seen the increasing importance of international trade, with attendant changes in economic, political and normative relations of global hegemonies. Contemporary China is perhaps the best illustrative of this transformation. In the last two decades we have witnessed the rise of China as a new economic super power, with the emergence of a Chinese middle class numbering hundreds of millions, and new international divisions of labour and consumption where China has been called the 'industrial area of the world' (Zhang & Ong 2008). In parallel with such developments has been a lively debate on the sustainability of international trade; crystallised most recently in the Rio +20 'Outcomes Document: The Future we want,' where world leaders endorsed calls for '...sustainable patterns of consumption and production...' (United Nations 2012, Paragraph 4 & Paragraphs 224-226). Sustainability can be approached from many perspectives[135] though this chapter takes as its point of departure notions of *sustainability as intra-generational equity*, and sustainable trade that equally has regard for the needs and aspirations of consumers and producers. Framed such, sustainability questions are ethical questions, asking what it is that consumers and producers value, and what principles they endorse for their society; things we introduce in this chapter as mapping an 'ethical terrain.' Given the rise of China as a central trading partner globally, sustainable trade demands an appreciation of the different perspectives of the multitudinous Chinese stakeholders engaged in this trade; what are the values and principles that comprise the diverse 'ethical terrains' of China? What is it that different Chinese producers want from this trade? It is to these questions that this chapter turns with a focus on the international trade of aquaculture products, and drawing broadly on the authors' multi-disciplinary perspectives and tools from applied ethics, political philosophy, and sociology. More specifically, this chapter presents research into what comprises the ethical terrain of the Chinese aquaculture sector.

[135] For some relevant discussions, see Stephanson & Mascia 2009, Kates *et al.* 2005, De Vries & Petersen 2009; for a discussion of the value-infectedness of sustainability and precaution in aquaculture, see Kaiser 1997.

C. Brautaset and C.M. Dent (eds.) **THE GREAT DIVERSITY**
DOI 10.3920/978-90-8686-786-8_10, © Wageningen Academic Publishers 2014

This chapter presents some findings of the on-going EU-funded research project, 'Sustaining Ethical Aquaculture Trade' (SEAT), which seeks to describe the trade of aquaculture products from Asia to Europe[136], and prescribe indicators of ethical and sustainable trade, with a particular focus on China, Bangladesh, Thailand and Vietnam, comprising the production of four species: shrimps, prawns, tilapia and pangasius. As part of the SEAT Project, the ethical terrain was mapped for those engaged with the Chinese aquaculture sector around Maoming, in southern China. The chapter begins in the second section with some background on contemporary developments in China and their broad influence on the ethical terrains of China more generally, before the third section that focuses on previous work done in the aquaculture sector. The fourth section introduces the SEAT Project and the methods used for the research, before the fifth section that outlines some specific results of the research. The sixth section presents discussion and conclusions on the implications of this research for sustainable aquaculture trade, with reference to the broader Chinese context and international certification schemes. It is important to explore how China's 'rise' has reshaped its ethical landscape in order to understand the cultural, political and economic exchanges between China and the outside world, particularly concerning questions related to sustainable trade in an increasing globalised economy.

From 'hard work and plain living' to 'realising one's dream'

Which key historical events have shaped contemporary Chinese normative discourse, and the ethical landscape? The revolution of 1911 and the end of the imperial dynasties, the Japanese invasion, the founding of the People's Republic of China in 1949, the Cultural revolution and the Reform and opening politics (*Gaige kaifang*) which started in 1978 are some of the most central events in the recent history of China. Historical experiences of war and revolution have elevated social stability as a central value, dominating Chinese political practice. The status and importance of social order and a 'harmonious society' can be seen in this context, with this similarly reflected in evolutions in the religious landscape (see below). For instance questions of economic polarisation, the balance between economic growth and the environment, and the relationship between the market and state are usually seen in light of a concept of harmony (Hansen & Thøgersen 2008). This is also expressed at the governmental level in that Hu Jintao has taken the concept of a 'harmonic society' as the point of departure for his leadership; something we find again in both the 11th and 12th five year plans.

[136] For an overview of EU policy on aquaculture and aquaculture trade, cf. EC COM 2009.

This much noted, mapping the religious landscape and related 'ethical landscape' in China is very difficult. This is both because China is a multi-religious country, and because the question of what should count as religion is perpetually debated (Kipnis 2001). Under Mao's leadership, the Chinese Communist Party either eliminated or brought religious practices under Marxist guidance. Religion became perceived as mere 'superstition' or 'bourgeois ideology', which partly explains why many Chinese are unlikely to identify themselves as religious. In spite of this, religious rituals and traditions have been revived after the liberalization of policies concerning religion since 1982 (Overmyer 2003). The government has also realized that religion can be an integrated element in the process of rebuilding a new China (Angelskår 2011), though many political commentators today will emphasise how Confucianism is more of a rhetoric feature of government rather than as a proponent of a philosophical outlook as such.

According to Kipnis (2001), three 'systems of thought', namely Confucianism, Taoism and Buddhism, could be described as constituting the complexity of Chinese culture, and some core values. First, the Chinese are preoccupied with reputation. Consequently, social cohesion and order is managed and maintained through feelings of shame. Second, the Confucian ideal of preserving harmony is deeply embedded in the Chinese society, with the thought that this ideal can be attained through moral conduct, controlling emotions and avoiding conflicts. Third, trust is fundamental in all social relations; in China, any weaknesses in the formal legal institutional framework are addressed through developing personal networks. These traditional values have in more recent years been challenged by new trends towards individuation and a highly materialistic orientation (Kipnis 2001).

There has been an apparent change from collectivism as the dominant socioeconomic ethos during the Mao-era, to a more individualistic morality where self-interested pursuits and the expression of personal desires are more acceptable (Westad 2012). These changes are seen to follow the reform and opening of politics and the transition from a system of planned economy to a late modern market economy. To talk directly about one's self-interest and well-being is no longer regarded as vulgar and politically incorrect (Yan 2009). Nevertheless, it is an individualism that has retained some of the traditional collectivist aspects, in which for instance loyalty to the family and the community is still a strongly held value. Indeed, it is argued that some of these traditional collectivistic aspects may be limiting women's mobility, and it is consequently not unusual that agricultural (or aquaculture) labour falls in the domain of women, as a result of men migrating to the cities for work (Brugere *et al.* 2001, Song 1999).

The shift away from collective moral experience of responsibility and self-sacrifice to a more individualistic morality that emphasises rights and self-cultivation must also be seen in the context of consumerism and new forms of liberal disciplinarian processes. With the politics of reform and the following de-collectivisation, the government in parallel with the market economy has instigated the institutional untying of the individual. For instance, the government and the mass media encouraged people to take out loans for commodities and services, through the political campaign named 'borrowing money to realise one's dream'. This marked a sharp contrast to the communist ideology of the pre-reform period (1949-1978), when 'hard work and plain living' was promoted as the ideal (Yan 2009, 2011).

Nationalism is also seen as a central collectivistic, political and normative imaginary for the Chinese today (see also Medvedeva's chapter in this book). This is a process is at significantly attributed to state initiated campaigns and the revitalisation of Confucianism and other religious practices. It is also influenced by events of the last decade, such as the bombing of the Chinese embassy in Beograd in 1999, American interference in China's relationship with Taiwan, the emergency landing of an American spy plane on Chinese territory in 2001 and historical experiences, especially from the period after the Republic was established. Such nationalism can explain the relatively large support for the Communist Party (CPC). From the end of the 1980s and beginning of the 1990s, there is a tendency that the party has gone from being Maoist to gradually developing a Chinese form of nationalistic ideology (see among others Unger and Barmé 1996). This is some of the cultural, economic and historical context for mapping the ethical terrain of China today.

Previous research on the 'ethical terrain' of the Chinese aquaculture sector

Although China has been prominent in research on aquaculture in a global context, a focussed study on stakeholders' normative terrain in this industry has not yet been fully realised. The academic literature on the development and sustainability of the aquatic food industry in Asia has accumulated in proportion to the expansion of the industry since the 1970s (Brugere *et al.* 2001, Song 1999). However, previous research on values related to the Chinese aquaculture industry in particular has primarily focused on gender issues and religion. In China, women account for more than one third of the total labour force in aquaculture (Song 1999), though there is not much literature on women's specific involvement. The existing literature seems to suggest the prevalence of gendered agricultural labour (including aquaculture) whereby it is not unusual that agricultural labour (including aquaculture) falls in the domain of women, as a result of men migrating to the cities for work (Song 1999). The absence of men elevates women's decision-making and independence. However, this independence of women in the aquaculture sector has been argued, by Brugere *et al.* (2001) for instance, to come into conflict with pervading Confucian ideologies, which limit women's mobility and status. This represents a very complex interaction between gender and religion (for a more comprehensive discussion on this topic, see the discussion by Xiaojiang 1994). In the absence of comprehensive research into the values and ethical principles comprising the ethical terrain of Chinese aquaculture sector, the SEAT Project undertook empirical work over 2010 and 2011.

Empirical research: the SEAT Project and its methods

Sustaining Ethical Aquaculture Trade

SEAT is a four-year European Union 7th Framework-funded research project initiated in 2009 to explore the trade of farmed seafood from Asia to Europe. It brings together 13 partner institutes from across Asia and Europe with the joint goals of (a) *describing* the 'value-chains' of aquaculture products stretching from Asia to Europe, and (b) initiating a *normative* discussion on how these value-chains ought to be more ethical and sustainable. SEAT explores multiple value-chains related to four different species in four different countries, but this chapter will limit its discussion to the particular chains of shrimps, tilapia, and the combined cultivation of tilapia and carps in China.

SEATs normative aspirations are steered by an 'applied ethics' work-package, led by the University of Bergen. Importantly, SEAT does not apply 'grand ethical theories' of sustainability to the value-chains; rather, by beginning from the heuristic of a complex value-chain, the project highlights the plurality of stakeholders interacting with the chain from farm to fork, and their disparate perspectives on what constitutes ethical and sustainable seafood trade. This point of departure brings to the fore notions of sustainability in terms of 'intra-generational equity' and 'justice' (see for example the discussion of Okereke 2006 on intra-generational sustainability, or Throsby's 1995 discussion of linking economic and cultural systems for intra-generational sustainability), such that sustainable trade recognises consumers and producers inhabit very different social, cultural and economic realities, and possess different values, principles and conceptions of 'the good life'. It is important that sustainable trade has fair regard for the well-being and aspirations of stakeholders' along the value-chain, while also attending to the more often evoked notions of sustainability as 'inter-generational equity'; '*...without compromising the ability of future generations to meet their own needs*' (intergenerational equity finds a greater resonance in the often cited report by the World Commission on Environment and Development 1987).

Towards its normative goals, SEAT is assembling a toolbox of indicators of ethical and sustainable aquaculture that combines the 'top-down' expertise of scientists and experts, with the 'bottom-up' realities of actors along the Chinese aquaculture value-chain, and those indicators that they express as being salient, legitimate and credible for them. To assemble this toolbox SEAT sought first to map the values and principles important to actors along the value-chain; discussed in terms of mapping the diverse 'ethical terrains' from China to Europe. These maps of the ethical terrain provided the background, or normative substance, that underpinned and gave weight to indicators within the framework of sustainability. Put in another way, indicators were treated as practical expressions of those things that consumers and producers value as important. A snapshot of the ethical terrain of European seafood consumers was pieced together via desktop review of quantitative studies with different groups of European consumers between 2000 and 2011. In China, SEAT research sought to piece together one particular perspective on the ethical landscape of stakeholders of the aquaculture sector in southern China, around Maoming. In this chapter, this is referred to as the 'ethical terrain of Chinese aquaculture'.

Methods and data

Mapping of the ethical terrain of Chinese aquaculture progressed according to a two-stage method. The first stage was undertaken in autumn 2010 as part of a quantitative 'integrated survey' of 405 Chinese aquaculture farm owners and managers around Maoming, in Guangdong province (376 males and 29 females), designed by University of Bergen partners and carried out by project partners at Shanghai Ocean University. In face-to-face interviews approximately one hour long, farmers were asked about all aspects of their aquaculture practice, including a short section asking respondents to rank two pre-prepared sets of eight values in terms of their importance to them. This value-ranking exercise constituted the first input for mapping the ethical terrain of Chinese aquaculture. The first set unpacked those values respondents saw as important for their personal concept of the 'good life', while the second set explored those values respondents saw as important foundations for a 'good society;' respectively offering insights into values important to the individual and to the collective (see 5th section *Value system 1: individual values* and *Value system 2: societal values*). Respondents were presented with the two sets of eight pictorial cards (16 in total), each representing a value, and asked to rank the values in their respective sets by placing the cards in two columns, with the most important values at the top down to the least important at the bottom. In a pilot study the two sets of values were formulated in partnership with the Chinese partners. A translation was made of the English word for the value into the local language, and then by way of a quality check, was re-translated back into English by an independent translator. Beyond translation, each value card featured a picture, or collage of pictures, representative of that value.

The rationale for grouping the values in these two sets is to grasp the values related to aquaculture both from a collectivistic and individualistic perspective. To this extent we included different indicators on values that communicate ethical standings and value based judgments related to the society as a whole, and values connected to individual livelihood issues. Of course such a clear categorisation of something as abstract and ephemeral as values is in some sense highly artificial, since such values are unlikely to be mutually exclusive, but it did provide a useful heuristic for our mapping exercise.

The 2nd stage was undertaken in spring 2011, as a qualitative follow-up to the integrated survey, to give substance to the ethical terrain emerging from the first stage by engaging a greater diversity of stakeholders, and asking them about management principles they considered important to the ethical and sustainable operation of the aquaculture sector. 11 face-to-face interviews were carried out with stakeholders from along the aquaculture value chain, and from within the local community; including government administration, hatchery workers, aquaculture farmers, input-suppliers, local residents and retailers, and local teachers for example. Stakeholders were determined to be both those who were either directly or indirectly affected by the aquaculture sector, and who demonstrated knowledge of the sector. Interviewees comprised seven males and four females ranging from 16 to 48 years old, though most (eight) interviewees were in their 30s or 40s. Interviewees were either recommended by SEAT Project fieldworkers, or were approached in the field, and were selected according to a basic imperative to access as many diverse perspectives on aquaculture as possible.

Interviews were undertaken by a SEAT Project partner from the University of Bergen, together with a local translator from Shanghai Ocean University, though were largely conducted in the local language; lasting for 20-35 minutes. The interview protocol itself was divided into two sections, one exploring the desirability of ethical management principles for respondents, and the other an exercise in scenario analysis; the latter extending the scope of this chapter and thus not discussed here. Respondents were introduced to nine principles through a fictitious narrative about river management, which paused at important junctures and asked respondents about the desirability of a proposed management principle, such as the Polluter-Pays Principle or the Precautionary Principle for example, using a five-point Likert scale. The Likert scale asked respondents to specify their agreement or disagreement with each principle on a symmetric 'agree-disagree' scale, such that the range captures the intensity of their feelings for each principle.[137] There are a number of advantages and disadvantages to the Likert-scale-type survey. In this study, the main advantage was that it was a quick and clean means of response, with interviews in the field often complicated by respondents needing to work at the same time, or being conducted on a roadside with crowds of bystanders, for instance. The main disadvantage is that a respondent's complex feelings regarding something so fundamental as key management principles is simplified to five increments of tick-boxes. This can skew results (because respondents are not clear if they find a principle a 'little' or 'very' desirable), and does not enable a deep exploration of why respondents feel as they do.

The ethical terrain of the Chinese aquaculture sector: key findings

In mapping the ethical terrain of stakeholders in the Chinese aquaculture sector, we began by unpacking the values important to stakeholders, dividing them into the two sections, *Value System 1 on individual values* and *Value System 2 societal values.* The frequencies are displayed in the Tables 10.1 and 10.2.[138]

Value system 1: individual values

The empirical data is based on the rankings of a pre-selected set of values aimed at a non-randomized selection of respondents employed in different roles in the Chinese aquaculture industry. By exploring what we term a 'landscape' or 'terrain' of the selected values with descriptive statistics as basis. However, this approach may also hold various implications. First and foremost, we intend to demonstrate the crude distribution of the given values and how these distributions *theoretically* can be said to correspond with the wider cultural and economic context in the Chinese society (these reflections on the validity of the data adds well to the analysis). The data is not to be interpreted as representative for the Chinese population as a whole, neither to be generalized as highest ranked values among the Asian aquaculture industry in general. Furthermore, we find it important to stress that the descriptive statistics used in the following discussion does not arrive from *a priori* hypothesis, and must

[137] The five-point Likert Scale gave respondents five possible choices relative to each principle: 1. Very desirable; 2. A little desirable; 3. Neither desirable nor undesirable; 4. A little undesirable; or 5. Very undesirable.

[138] Tables 10.1 and 10.2 only display the distribution of the highest ranked values in terms of the corresponding frequencies. In the analysis, we operated with a three-fold division, separating between the 1st, 2nd, and 3rd rank respectively. Here we find similar patterns to that of the 1st ranked values among the respondents asked in the survey. *Family and household* (31.9%), *personal prosperity, happiness and well-being* (17%), and the value of *future generations* (14.1%), compromises the 2nd highest ranked values. This is somewhat similar to the values ranked the 3rd highest, constituted by *personal prosperity, happiness and well-being* 18.5%), *friends and relatives* (18.5%), and *family and household* (15.8%).

therefore not be regarded as statistically significant findings in the conventional meaning of the term.

As Table 10.1 shows, there is a clearly skewed distribution among the respondents in terms of the highest ranked values in this first section of variables. Roughly four individual value indicators seem to predominate among the Chinese stakeholders interviewed. In descending order, we find that 'country' is the highest ranked value by approximately 30% (N=123) of the respondents asked in the survey, followed by 'family and household' by 28% (N=112), 'personal prosperity, happiness and well-being' by 20% of the respondents (N=80), and furthermore, the value of 'Future generations of community and country' comprised by roughly 8% of the respondents (N=31). On the other hand, we can see the value of 'friends and relatives', 'animals and nature', 'religion and God'[139], and the value of 'local community', as approximately 5%, 1.5%, 1.2%, and 1% of the respondents respectively. But how can the pattern of values be said to correspond to the values in the second section or values regarding a good society?

Table 10.1. Chinese aquaculture farmers' individual values for 'the good life' (1st rank).

	Frequency	%	Valid %	Cumulative %
Country	123	30.4	30.4	30.4
Family and household	112	27.7	27.7	58.1
Personal prosperity, happiness and well-being	80	19.8	19.8	77.9
Future generations of community and country	31	7.7	7.7	85.6
Friends and relatives	19	4.7	4.7	90.3
Animals and nature	6	1.5	1.5	91.8
Religion and God	5	1.2	1.2	93.0
Local community	4	1.0	1.0	94.0
Total	380	94.0	94.0	

[139] The categories of values was standardised across all four Asian countries studied by the SEAT Project for reasons of comparison. However, this was particularly problematic for the value of 'religion and God', with very different religious belief systems across Bangladesh, China, Vietnam and Thailand, and indeed 'religion' understood in different ways. In China, the value card for religion and God was illustrated by a photo of a statue of Buddha, as was supposed to be symbol of organised belief systems.

Value system 2: societal values

By turning to the 2nd section, we find a somewhat more dispersed distribution. Although the ranking is spread more evenly, we can still identify the highest ranked values among the respondents asked in the survey, with the 'top four' values comprising of 'health, safety and income security' by 38.5% (N=156), the value of 'social harmony' by 16% of the respondents (N=65), 'individual freedom to such as speech, vote and religion', shared by approximately 9.5% (N=38), and 'environmental concerns' with just 8% (N=32). 'Respect between humans' is not a value given the highest priority either, shared by only 7% of the respondents, closely followed by the value of a 'strong political leadership' by 6% approximately. Furthermore, we find that only 5.2% share the value of 'equal personal opportunities such as gender, age and class'. Least priority is given to the value of 'collaboration and cooperation in community' as shown in Table 10.2.

Table 10.2. Chinese aquaculture farmers' values for 'a good society' (1st rank).

	Frequency	%	Valid %	Cumulative %
Health, safety and income security	156	38.5	38.5	38.5
Social harmony	65	16.0	16.0	54.5
Individual freedom to such as speech, vote, religion	38	9.4	9.4	63.9
Environmental concerns	32	7.9	7.9	71.8
Respect between humans	31	7.7	7.7	79.5
Strong political leadership	25	6.2	6.2	85.7
Equal personal opportunities such as gender, age, class	22	5.4	5.4	91.1
Collaboration and cooperation in community	12	3.0	3.0	94.1
Total	381	94.1	94.1	

Management principles contributing to the ethical terrain

In the 2nd stage of the research, 11 key-informants were presented with a narrative that illuminated nine ethical principles,[140] and asked to comment on their desirability using a five-point Likert scale. Such a small sample size does not offer the same robustness as the sample used for the value-ranking exercise, and therefore the results should be viewed with caution. However, it does augment the value-ranking exercise by offering some very broad signals on ethical principles seen as desirable by some Chinese aquaculture stakeholders, and thus offering a qualitative follow-up to interpret the quantitative data. Of the nine ethical principles interrogated, the final question on *solidarity* was determined to be imprecise, such that the discussion is limited to the first eight principles (identifying both methodological strengths and limitations – as you do here – makes both the approach and arguments more convincing).

The 'polluter-pays'[141] and 'prevention'[142] principles were most widely seen as desirable by those interviewed, with unanimous support for the polluter-pays principle. Those interviewed voiced some emphasis on self-responsibility, and clear ideas of where responsibility should lie. Associated with the prevention principle, the 'precautionary' principle[143] was also seen by most as desirable though less so than prevention.[144] Most of those interviewed saw principles supporting 'equal access to basic needs'[145] and 'equal working conditions'[146] as desirable. However, this support was mixed, with local retailers and residents and government administrators not finding either of the principles desirable. This is interesting because it may indicate a more complex personal relationship to notions of equal working and living conditions than may be endorsed collectively.

140 These nine principles, in order of appearance in the narrative were: 1. The polluter-pays principle; 2. The prevention principle; 3. The precautionary principle; 4. Equal access to basic human needs; 5. Respect for nature; 6. Respect for minorities; 7. Equal working conditions; 8. Respect for animal welfare; and 9. Solidarity.

141 The principle states in effect that the source of pollution should also be responsible for paying the clean-up.

142 This principle states that sometimes a polluter may have to take preventive action even when the documented effect only affects a very small proportion of the population.

143 This principle, being different from the prevention principle, states that one should act preventively even when the possible harm is uncertain and only suspected by indirect evidence.

144 Interestingly, all four of the female interviewees saw the precautionary principle as undesirable. Whether this is indicative of gender-difference in approaches to risk, or simply resulting from a small-sample size is unclear. On the other hand, it may be indicative of interviewees' understanding of the principles of prevention and precaution, with prevention potentially more easily grasped than precaution. It may also be due to the fact that prevention was exemplified by reference to harm to pregnancies while precaution was exemplified by reference to harm to the environment.

145 In the study, this principle was exemplified by equal prices for drinking water despite different accessibility.

146 In the study, this principle was exemplified by minimum wages for all workers, regardless of their origin e.g. immigrant workers.

The principles of 'respect for nature'[147], 'respect for animal welfare'[148] and 'respect for minorities'[149] were seen as the least desirable among those interviewed. Respect for nature was seen as desirable by 6 of 11 respondents, and this principle also saw the most interviewees register a neutral opinion, or that they did not know. Both the respondents who noted respect for nature as undesirable were female aquaculture farmers. On the other hand, a majority of those interviewed saw respect for animal welfare as an undesirable principle. Finally, the principle of respect for minorities was seen as the least desirable of all principles, with only three respondents registering it as desirable.

Discussion and conclusion

From collectivism to individualism?

The findings of the empirical work appears to resonate with the more general reflections on the Chinese ethical terrain nationally, captured in the broad and diverse literature touched on in the 2nd section. The distribution of the various frequencies indicate that 'collectivistic' values of country and family and household in particular hold a strong position among the Chinese interviewees. In the rural areas around Maoming, the family remains an important social institution and a source of social security, stability and harmony in which individuals' needs are, in large, subordinate to family and group interests. These areas have retained primary production, and with that, income tied to the family's common agricultural enterprise (Hansen & Thøgersen 2008). To an extent we see this extended also in the relatively high value placed on 'future generations', which could also be interpreted as regard for those who will in future inherit the family farm. Both the values 'country' and 'family and household' reflect a stronger cultural complexity rooted in an industrial ethos in the Chinese contemporary, late-modern economic structure. On the other hand the results revealed, in view of the traditional Chinese values, a tendency toward individualism; that is, placing value on 'personal prosperity, happiness and well-being' (ranked 3rd in personal values) but little on 'local community,' and even more markedly, placing value on social values that protect the rights of individuals to realise their aspirations and self-cultivation, notably on 'health, safety and income security' and 'personal freedom,' though putting much less value on 'collaboration

[147] In the study, this principle was exemplified by a dispute about the protection of a waterfall against the interest of energy production.

[148] In the study, this principle was exemplified by the claim that local pig farmers should allow minimal spaces for the pigs according to international standards.

[149] In the study, the principle was exemplified by the alleged protection of a lake as a religious site for a religious minority.

and cooperation in the community.' These observations raise both ethical and value related issues in terms of aquaculture trade in a broader global perspective and how it organises and affects the life chances and social economic positioning for the stakeholders involved.[150]

The ethical terrain of Chinese aquaculture indicates the unique intertwining of individual and collectivist values. Though individualistic values have gained acceptance and to a large degree have been enforced and embraced to position China on the global economic stage, the predominance of collectivistic values still occupies an important role connecting Chinese aquaculture products with the European market. *Per se* this trend can be said to reflect the current political and economic outlook in China, where the Party Communist ideology, in many ways, is driven by strong global market incentives.

A third strong influence on the ethical terrain of Chinese interviewees is that of nationalism, as discussed in the 2nd section. We see that respondents overwhelmingly saw their country as that which they personally value the most. Of course, this data, as for all data resulting from this empirical research, demands a critical re-examination into the degree that interviewees responded truthfully or as they felt they were expected to respond. As an on-going dilemma facing the social sciences, it is well-documented that the interviewer and social context within which an interview is conducted can influence an interviewee's answers. In a country where nationalism is endorsed as an ideology and popular movement (see 2nd section), it could be anticipated that interviewees felt a pressure to respond as they did. The only safe-guard against this is well-designed interview protocols and well-trained interviewers for eliciting the most truthful responses possible.

Two other notable observations can be made of the value ranking exercise. The first is consistent with a long period of communist rule and describes the reduced value placed on 'religion' by individuals, to be replaced by a reverence for the state. While other studies show that religion retains an important role in Chinese society (Shijun 2009), and we saw in the 2nd section the re-emergence of Confucianism, this study with Chinese aquaculture stakeholders seemed to indicate a reduced regard for religion. As another critical reflection, it is important to note the possible influence of using a simplistic value-card for 'Religion and God', which may have meant respondents had a narrow or misplaced notion of what was meant by religion, that did not correspond to the broad interpretation that was meant. The other interesting

[150] Among the values ranked 2nd, we located respect between humans (19.3%), health, safety and income security (18%), followed by the value of social harmony (12.8%). The values ranked 3rd highest is constituted by respect between humans (18%), equal personal opportunities (16.3%), in addition to social harmony (13.8%).

observation is relative to the relevance of 'sustainability' for the Chinese aquaculture sector, with the relative high value placed by individuals on 'future generations', and recognition that a good society should value its environment. This being noted, individuals placed little personal value on animals and nature.

Beyond values, we explored the ethical principles that those in the aquaculture sector saw as important for steering the sector. While recognising the very small sample size, there were some interesting broad patterns that emerged. Above all was the support for the 'polluter pays' principle, and the attendant endorsement of personal responsibility. This could correspond with the rise in individualism, and by extension, backing for ethical principles of autonomy where individuals are left the freedom to pursue their own ends, provided they do not infringe on others. Principles of prevention and precaution also found strong support, possibly linked to the relative value put on sustainability discussed above, though in strong contrast to the unbound economic growth seen in the aquaculture sector, and Chinese industry more generally, over the past decades. At the other end of the scale, respondents were seen to show little support for principles that demand respect for the weak or the voiceless; notably minority groups, animals or the natural environment. Particularly the principle of 'respect for minorities' was roundly rejected, which seems to indicate a preference for advancing the interests of the majority; congruent with the personal value put on the country and the social value put on social stability and a harmonious society.

Regard for the Chinese ethical terrain in global trade

In this final juncture we return to the key message at the beginning of the chapter, arguing that in the context of global trade producers and consumers ought not to exist in mutual exclusivity. In the interests of intra-generational sustainability, trade must look beyond the simple satisfaction of consumer demand to frame trade as a 'value-chain', with an appreciation for the well-being of stakeholders along the length of this chain. For instance, this implies a shift from asymmetrical, consumer-focussed trade-certification schemes, to certification that recognises the well-being of all interest groups, from producers to consumers. Using the heuristic of the chain, we can see how global trade has influenced the ethical terrain of China; bringing with it an ethic of individual responsibility, autonomy, personal freedom and wellbeing that is congruent with a free-market globalised economy, and introducing a certain degree of conflict with the pervading communalist and, indeed, communist ideologies of China. However, having noted the role of global trade on the Chinese ethical terrain we can pose the question, what influence has the Chinese ethical terrain on global trade? And what role ought it to have? It should also be said that discussions about the normative validity of either the European or Chinese value system remain, of course, unaffected by this descriptive study.

This chapter presents a modest step towards recognising what stakeholders in the Chinese aquaculture sector value, and what ethical principles they place above others for steering the sector and the trade. Such reflexivity and the learning that follows should influence the shape of trade in the future, by recognising those things we place common value on and other things where we diverge. By way of example, our study appeared to demonstrate a relative value put on sustainability and precaution in by our respondents in China commensurate with the support we see among seafood consumers in Europe for example, while on the other hand, the respondents showed greater 'nationalism', but less regard for the natural environment and animals, than European consumers (see e.g. the Eurobarometer surveys; European Opinion Research Group 2003; European Commission 2003, 2006). Common understanding among trading partners is a first step to sustainable trade with regard for intra-generational equity.

Acknowledgements

The authors would like to acknowledge the European Union 7th Framework Programme for funding the SEAT Project that made this research possible and the SEAT Project consortium for all of their support. Particular acknowledgment must go to our SEAT Project partners at Shanghai Ocean University, without whom this fieldwork could not have been possible. A special thank you to Wenbo Zhang, Huanan Liu and Liping Liu for their tireless work in the field.

Bibliography

Angelskår, T. 2011. Kinas nye opium. *Innsikt*. Available at: http://www.aftenposteninnsikt.no/11-desember-2011/kinas-nye-opium.

Brugere, C., Kusakabe, K., Kelkar, G., Felsing, M. & Muir, J. 2001. *Women in Aquaculture*. Project FWG 03/99, Asia Pacific Economic Cooperation, Singapore.

De Vries, B.J.M. & Petersen, A.C. 2009. Conceptualising sustainable development: An assessment methodology connecting values, knowledge, worldviews and scenarios. *Ecological Economics* 68(4): 1006-1019.

European Commission. 2006. *Special Eurobarometer 246: Health and Food*. European Commission, Paris, France.

European Commission Communications (EC COM). 2009. *Building a sustainable future for aquaculture – A new impetus for the strategy for the sustainable development of European Aquaculture*, Communication from the European Commission to the European Parliament and the Council, COM (2009) 162 final, 8.4.2009.

European Opinion Research Group. 2003. *Special Eurobarometer 186: Health, Food and Alcohol, and Safety*. European Commission, Paris, France.

Hansen, M.H. & Thøgersen, S. 2008. *Kina – individ og samfunn*. Universitetsforlaget, Oslo, Norway.

Kaiser, M. 1997. Fish-Farming and the Precautionary Principle: Context and Values in Environmental Science for Policy. *Foundations of Science* 2: 307-341.

Kates, R.W., Parris, T.M. & Leiserowits, A.A. 2005. What is sustainable development? Goals, indicators, values and practice. *Environment: Science and Policy for Sustainable Development* 47(3): 8-21.

Kipnis, A.B. 2001. The Flourishing of Religion in Post-Mao China and the Anthropological Category of Religion. *The Australian Journal of Anthropology*. 12(1): 32-46

Okereke, C. 2006. Global environmental sustainability: Intragenerational equity and conceptions of justice in multilateral environmental regimes. *Geoforum* 37(5): 725-738.

Overmyer, D.L. 2003. Religion in China Today: Introduction. In: *The China Quarterly*. Cambridge University Press, Cambridge, UK.

Shijun, T. 2009. *Spiritual Life of the Contemporary Chinese Population*. Economic Sciences Press, Beijing, China.

Song, S. 1999. *Rural aquaculture in China*. RAP Publication 1999-22, FAO Regional
Office for The Asia and Pacific, Bangkok, Thailand.

Stephanson, S.L. & Mascia, M.B. 2009. *Putting people on the map: An approach to integrating social data in conservation planning. SSWG Working Paper 1*. Society for Conservation Biology, Social Science Working Group, Washington, DC, USA.

Throsby, D. 1995. Culture, Economics and Sustainability. *Journal of Cultural Economics* 19(3): 199-206.

Unger, J. & Barmé, G. 1996. *Chinese Nationalism*. M.E. Sharpe, Armonk, New York, NY, USA.

United Nations. 2012. *Rio +20 Outcome Document: The Future We Want*. The United Nations, Rio De Janeiro, Brazil.

Westad, O.A. 2012. China's geoeconomic strategy: China's international future. In: Kitchen, Nicholas (ed.). *IDEAS reports-special reports, SR012*. LSE IDEAS, The London School of Economics and Political Science, London, UK.

World Commission on Environment and Development. 1987. *Our Common Future*. Oxford University Press, Oxford, UK.

Xiaojiang, L. 1994. Economic reform and the awakening of Chinese women's collective consciousness. In: Gilmartin, C. (ed.). *Engendering China: Women, Culture, and the State*. Harvard University Press, Cambridge, MA, USA.

Yan, Y. 2009. *The Individualisation of Chinese Society*. London School of Economics monographs on social anthropology. Bloomsbury Academic, 2009, London, UK.

Yan, Y. 2011. The Changing Moral Landscape. In: Kleinman A., Yan, Y., Jun, J., Lee, S., Zhang, E., Tianshu, P., Fei, W. & Junhua, G. (eds.). *Deep China: the moral life of the person. What anthropology and psychiatry tell us about China today.* University of California Press, Berkeley, CA, USA.

Zhang, L & Ong, A. 2008. *Privatising China: Socialism from afar.* Cornell University Press, London, UK.

Chapter 11: The growth trajectory of Indian multinational enterprises

Surender Munjal, Peter J. Buckley, Peter Enderwick and Nicolas Forsans

Indian multinational enterprises (MNEs) are rapidly growing in the world economy. This chapter explores the growth trajectory of Indian MNEs from an international business perspective. It investigates the growth of Indian MNEs in relation to the liberalisation of the Indian economy. Since liberalisation has been used as a tool for growth by many developing countries in Asia (Greenaway *et al.* 2002, Khanna & Palepu 2010), the growth trajectory revealed in this chapter may also be applicable to MNEs originating from other developing countries in Asia. The growth trajectory explored in this chapter is split into two periods: 1) growth during the pre-liberalisation period; and 2) growth during the post-liberalisation period.

It is explained that where the pre-liberalisation period has given Indian MNEs a protection from competition posed by foreign MNEs; the post-liberalisation period has given them a freedom to benefit from foreign MNEs. Indian MNEs seem to have exploited the freedom by acquiring and partnering with foreign firms. The post-liberalisation growth strategies of Indian MNEs are explored by using two theoretical frameworks put forward in the literature on international business. These theoretical frameworks are: 1) the Linkage-Leverage-Learning framework; and 2) the asset augmentation strategies; both of which together put forward a process and strategy of internationalisation applicable to MNEs from a developing country.

Linkage enabled Indian MNEs to learn while accessing strategic resources from foreign MNEs. Acquisition enabled Indian MNEs to swiftly establish its presence in the world economy while augmenting competitive advantages. As per an estimate, more than 70% of the outward foreign direct investment (FDI) by Indian MNEs is undertaken through acquisitions (Athukorala 2009). Different types of strategic assets, such as popular brands and technological assets are obtained through acquisitions and linkages, have enhanced the competitiveness of Indian MNEs in the global economy.

Using these strategies, many Indian MNEs have become leading firms in their respective industries; for example, Infosys and TCS in the information technology (IT) industry; Ranbaxy, Dr. Reddy's and the Chemical, Industrial & Pharmaceutical Laboratories (CIPLA) in the pharmaceutical industry; Suzlon in the wind power sector; the state owned Oil and Natural Gas Corporation (ONGC) in the energy

C. Brautaset and C.M. Dent (eds.) **THE GREAT DIVERSITY**
DOI 10.3920/978-90-8686-786-8_11, © Wageningen Academic Publishers 2014

industry; Tata Motors and Mahindra and Mahindra in the automobile industry; and Tata Steel in the steel industry. Of these cases, this chapter includes three illustrative cases – Suzlon, Tata Motors, and Hero Group. These cases provide a rich explanation of the evolution and growth trajectory of Indian MNEs.

Growth trajectory in the pre-liberalisation phase (1948-1991)

In the pre-liberalisation period, the Indian economy was heavily regulated and relatively closed. In fact, '...by the end of the 1970s, India had acquired a reputation as one of the most protected and heavily regulated economies in the world' (Kotwal *et al.* 2011: 1152). To a large extent, this was an outcome of India's experience of development with the help of foreign capital, provided by Britain, her colonial master. This made early Indian policy makers reluctant to rely on the inflow of FDI for the growth and development. Furthermore, India's first prime minster, Pandit Jawahar Lal Nehru, combined the principles of democratic socialism and economic self-sufficiency to build the nation in a way that avoided India's colonial past and dependence on Britain. In the post-independence pre-liberalisation period (1948-1991) the policies for economic growth adopted by the then government did not completely reject the role of inward FDI but kept it highly regulated (Balasubramanyam & Mahambare 2003).

Indian policy makers used trade tariffs and restrictions on inward foreign direct investment to protect and assist indigenous industries (Akamatsu 1962). Indian businesses were given monopolistic powers in the domestic economy because they were largely protected from global competition. In addition, industrial licensing and production quotas were also used as the main entry barrier to help local firms. Scholars, such as Pradhan (2007: 5) criticised the protective policy environment followed during the pre-liberalisation period, suggesting that the 'large sheltered' domestic market encouraged inefficiency and limited the export-intensity of Indian MNEs, 'as a result of which India's share in global trade was hovering below 1%' during the 1990's. Restrictions were also placed on the import of technology (Desai 1972); this further propelled inefficiency and prevented modernisation of Indian MNEs. However, a sheltered domestic market, particularly a large one like India's, offered an opportunity for local firms to serve the market in a monopolistic fashion. Due to the lack of competition, inefficiency did not become a barrier to growth. These firms were able to grow simply because they had ready access to the large domestic market. The effect was that many Indian firms generally remained profitable and over time accumulated huge financial reserves. Recent studies (Buckley *et al.* 2009, India Knowledge@Wharton 2006) have identified the importance of financial resources in explaining the rise Indian MNEs globally.

A further outcome of protectionist policies was the evolution of conglomerate business groups, such as Tata, Birla, and Ambani. These business groups are continuously dominating the Indian corporate world (PTI 2002) by expanding their scope of activities in different areas, from production of basic low technology goods, such as salt and textiles to high technology-intensive products, such as pharmaceuticals, telecommunications, software, aviation and petro chemicals. Further antecedents for the growth of Indian firms can be found from the explanations and theories in the international business literature, such as internalisation theory (Buckley & Casson 1976) and the role of market imperfection and institutional voids (Khanna & Palepu 2010). However, the focus of this chapter is not to discuss various antecedents of growth of Indian firms. Rather, the chapter focuses on the growth trajectory of Indian firms.

Growth trajectory in the post-liberalisation period (1991 onwards)

Contrary to the belief of many Indian policy-makers, the growth rates of Indian firms further accelerated with economic liberalisation. Particularly important was the liberalisation of India's outward foreign direct investment policy that allowed Indian MNEs to invest abroad (RBI 2012). In this period (1991 onwards), Indian MNEs started to use their financial resources including equity raised from the domestic capital market to undertake FDI much of which, perhaps 70%, was used for foreign acquisitions (Athukorala 2009).

In the international business and strategy literature, the phenomenon of internationalisation through acquisition is often associated with 'asset augmentation' strategies meaning that the MNE undertakes acquisition with a view to acquiring the strategic assets of the target firm to gain competitive advantages. Table 11.1 shows prominent changes in India's outward foreign direct investment regime and volumes of outward foreign direct investment. It can be observed from the table that liberalisation of the policy directly coincides with the rising level of foreign direct investment undertaken by Indian MNEs, at least until the global economic crisis in 2009.

Opening up of the Indian economy also allowed foreign MNEs to enter India. As a result, many local Indian firms started to face the challenges of global competitors. However, this also provided opportunities for Indian firms to grow by linking up with foreign MNEs. In this chapter, we do not discuss the challenges posed by foreign MNE in the form of rising competition as this is well documented elsewhere (for a detailed discussion see Porter 1986). Here we focus on how opening up of the Indian economy has provided opportunities for local Indian businesses to grow.

Table 11.1. India's outward FDI flows and policy changes (UNCTAD, FDI/TNC database. Available at: www.unctad.org/fdistatistics, Reserve Bank of India various bulletins, and Gopinath (2007)).

Year	Foreign direct investment (million US$)	Changes in outward foreign direct investment policy
1992	24.0	An automatic route for overseas investments was introduced and cash remittances were allowed for the first time. The total value was restricted to US$ 2 million.
1993	0.4	
1994	82.0	
1995	119.0	A fast track route was introduced where limits were raised from US$ 2 million to US$ 4 million and linked to average export earnings of the preceding three years.
1996	240.0	
1997	113.0	Indian promoters were allowed to set up second and subsequent generation companies, provided the first generation company was set up under the fast track route.
1998	47.0	
1999	80.0	
2000	514.4	The limit for investment up to US$ 50 million, which was earlier available in a block of three years, made available annually without any profitability condition. Companies were allowed to invest 100% of the proceeds of their depository receipts issues for acquisitions of foreign companies and direct investments in Joint Ventures (JVs) and Wholly Owned subsidiary (WOSs).
2001	1,397.4	
2002	1,678.0	Automatic route was further liberalised wherein Indian parties investing in JVs/WOSs outside India were permitted to invest an amount not exceeding US$ 100 million.
2003	1,875.8	Automatic route was significantly liberalised to enable Indian parties to fund to the extent of 100% of their net worth.
2004	2,175.4	
2005	2,985.5	The ceiling of investment by Indian entities was revised from 100% of net worth to 200% of the net worth of the investing company under the automatic route for overseas investment.
2006	14,285.0	
2007	19,594.4	The limit of 200% of the net worth of the Indian party was enhanced to 400% of the net worth under automatic route.
2008	19,256.5	Indian companies have been allowed to invest in excess of 400% of their net worth as on the date of the last audited balance sheet in the energy and natural resources sectors, such as oil, gas, coal and mineral ores.
2009	15,927.1	
2010	13,151.0	
2011	14,752.0	

With the opening of the Indian economy in July 1991, many foreign firms were attracted towards the Indian economy due to its large market size, and lower costs of production especially lower wages. The presence of inward investors in India provided an opportunity for many Indian firms to link up with foreign MNEs in different ways, such as being contractors, suppliers, joint venture partners, franchisees, collaborators and so on. These opportunities for linking up have proved to be significant in the trajectory of growth of local Indian firms (Buckley *et al.* 2012). Thus, we argue that the post-liberalisation trajectory of growth of Indian MNEs has been based on two types of strategies: (1) linkages with foreign MNEs; and (2) asset augmentation. Both of these are briefly explained in the sections below. After explaining these two phenomenon we look at examples of Indian MNEs to explain their growth with the help of arguments developed in this chapter.

The Linkage-Leverage-Learning framework

The Linkage-Leverage-Learning framework, also known as the LLL framework, was developed by Mathews (2002). The LLL framework is a strategic framework as it tends to guide the strategies of managers of the firm internationalising in pursuit of resources. The framework presents a sequential process of internationalisation by drawing on the three L's i.e. Linkages, Leverage and Learning. The strategic path for the firm's growth that builds on Linkage-Leverage-Learning actually starts after the foreign MNE is attracted by the location advantages of the developing country.

With globalisation of the world economy, MNEs are venturing out of their home country for various reasons prominently to find new markets and to reduce costs of production. In this move, they normally go to developing countries where markets are relatively unsaturated, surplus labour is available, and operating costs are significantly lower (Agarwal & Ramaswami 1992, Balasubramanyam & Mahambare 2003). Entry of foreign MNEs in a developing economy provides an opportunity for local firms to build linkages with them. Managers of the developing country firm need to strategically identify different ways in which they can build business relationships with foreign MNEs. This could be by being a contractor or subcontractor for a firm outsourcing production, or being a joint venture partner with a firm establishing a sales network in the developing country. These linkages can then become a medium through which the developing country MNE can access resources of the foreign MNE. Benefiting from resources of the foreign MNE is known as leverage in the framework. There are many ways in which the managers of local developing country firms can strategically leverage their growth by accessing resources of the foreign MNE. For instance, a firm undertaking manufacturing on behalf of the foreign MNE tends to access the technological know-how, and designs of the foreign firm, a joint venture partner may be exposed to the management and marketing practices of the foreign firm. Mathews (2006) argues that the developing country MNE needs to focus on ways to gain access to these resources and develop absorptive capacity to benefit; keeping in view that some of these resources may have constraints, such as imitability, transferability, and substitutability. He contends that the developing country MNE can view the world as full of resources and find itself in a position where it have nothing to lose but everything to gain by tapping the resources of foreign MNEs. On the other hand, foreign MNEs may fear loss of resources or knowledge when forming strategic alliances or joint ventures with developing country MNEs. However, the foreign multinational has to establish linkages with the developing country MNE to reduce the cost of operations, accessing cheap labour, and other resources owned or possessed by the developing country MNE which can enhance competitiveness of the foreign multinational.

Finally, according to the LLL framework, the managers of developing country firm should focus on the strategies to learn; for instance, learning by doing and imitating foreign MNEs. Learning different practices and techniques, such as organisational practices, marketing and production techniques can eventually help the developing country MNE to grow and compete globally. However, in this process it is assumed that the developing country firm has the capacity to learn.

Thus, in the LLL framework, building linkages is the starting point for the internationalisation of a developing country MNE. The idea is that the developing country MNE will be able to access the resources of other more established firms through linkages. Linkages are necessary for developing country MNEs as they may lack the resources required for internationalisation. Furthermore, in order to enter and compete in risky global markets it is necessary that the developing country MNE does not put all its resources at risk. In the internationalisation process linkages can be established, maintained and developed not only with the aim of accessing resources of foreign MNEs but also to reduce risks and uncertainties arising from lack of market intelligence or the costs of foreignness (Barkema *et al.* 1996, Mathews 2006, Pfeffer & Nowak 1976).

Asset augmentation strategies

To foster growth and to overcome resource limitations, particularly in a globalising world, developing country firms may also follow strategies of asset augmentation. This means that they seek to acquire strategic assets, both tangible and intangible, such as technological know-how, well known brands, and skilled human resources to increase their competitiveness. It is important to note that the asset augmentation strategy differs from leverage. The asset augmentation strategy involves buying the assets, while leverage involves tapping into the resources of others.

Asset augmentation strategies are not exclusive to developing country MNEs; any firm can undertake such strategies, however, it is increasingly seen that developing country MNEs tend to use them to acquire strategic assets that can compensate for their disadvantages.

Developing country MNEs, due to the business environment in which they are created and grow, generally lack some strategic assets that can enhance their competitiveness (Makino *et al.* 2002). Their home countries normally provide them advantage in performing low skilled and low value adding activities which offer limited growth opportunities. Lower levels of economic development at home pose a limitation on the local availability of strategic assets needed to compete globally, such as technological know-how, well known brands and skilled human resources. Thus, internationalisation of the developing country MNE is generally triggered by the desire to acquire such strategic assets (Luo & Tung 2007). This is one of the reasons why acquisitions by firms from developing economies targeted at developed country firms have risen (KPMG 2012, Stucchi 2012).

The move to acquire strategic assets also allows the developing country MNE firm to leapfrog and establish rapidly in the global economy (Athreye & Kapur 2009, Mathews 2002). A study by Luo and Tung (2007: 485) argues that emerging market MNEs use internationalisation as a 'springboard' in the trajectory of growth by 'acquiring foreign companies or their subunits that possess knowledge-based assets, such as sophisticated technologies or advanced manufacturing know-how'. MNEs from the newly industrialised countries also followed asset augmentation strategies in their internationalisation during the 1980s (Dunning *et al.* 2008, Narula & Nguyen 2011). However, in order to undertake asset augmentation, the firm should have access to significant financial resources which seems to be the case with Indian firms (Buckley *et al.*, 2012). Having discussed the LLL strategic framework and the asset augmentation strategies of growth of the firm, Table 11.2 compares them to highlight which strategy is when applicable.

However, a point to note is that asset augmentation and the linkage-leverage-learning strategies are not mutually exclusive. A firm can follow both of them simultaneously. This point can be seen in the short case studies that follows. Further, it is also important to note that the firm should have the capacity to absorb and learn in order to derive maximum benefit from the resources accessed or bought. In both cases the firm tends to acquire resources that may or may not be otherwise available

Table 11.2. Comparing the Linkage-Leverage-Learning (strategic) framework with the asset augmentation strategies (Authors' compilation).

LLL (strategic) framework	Asset augmentation strategies
Resources are accessed strategically	Resources are bought out
Relationships are the key to gain resource access	Financial resources are the key to gain resource access
Useful strategy for the firm that lacks financial resources	Generally useful for the firm having sufficient financial resources
A comparatively slow process	A comparatively fast process
Access and transfer of resources may be limited	Full access and complete transfer of resources may be achieved
Requires sufficient absorptive capacity	Requires effective organisational assimilation

in the market. The following section provides three short case studies to showcase Indian firms that have used state protection, asset augmentation strategies, and the strategies for linkages with foreign MNEs to grow successfully.

Internationalisation of Indian multinational enterprises: case studies

From the plethora of cases of Indian MNEs, this section showcases the growth of three prominent firms. These Indian MNEs, coming from different industries, have successfully grown in their own distinct ways. All three cases have rich information and together they portray the growth trajectory typical of firms originating from developing economies. The highlighted three Indian MNEs are: Suzlon, Tata Motors, and Hero Honda.

Suzlon

Suzlon is a world leading wind turbine manufacturer. According to the company website, Suzlon is the world's fifth largest wind turbine producer, in terms of cumulative installed capacity, with a workforce around 13,000 and a presence in 33 countries spread across six continents of the world. The company has about 7.6% share in the global wind energy market (Suzlon 2012). Suzlon was formed in 1990 by Mr Tulsi Tanti who divested his textile business to start the firm. He ventured into the power generation business because of power shortages in India. From the point of view of Khanna and Palepu (2010) this was an institutional void which became an antecedent for the formation of the business. However, the growth of the company is based on its linkages with firms from Germany. In its initial years, Suzlon accessed wind turbine generator technology through licences from Sudwind GmbH Windkrafttanlagen, a German firm, to set up a small power generating wind turbine (up to 0.35 MW) in India. After a couple of years, Suzlon absorbed Sudwind's technology and started to develop its own designs for producing larger (up to 2 MW) wind turbine generators (Ramamurti & Singh 2010). This shows that the firm not only leveraged access to foreign technology but also mastered it. In this case, the company grew by accessing foreign resources and learning from its linkages with a foreign MNE.

Later in 2001, Suzlon set up a Greenfield R&D unit called AE Rotors in the Netherlands and a year later, it acquired Sudwind, its earlier technology provider that was facing financial distress. These FDIs strengthen Suzlon's European R&D base. In 2006, Suzlon acquired Hansen, a Belgian gearbox manufacturer, for € 465 million and subsequently in 2007 Suzlon acquired another German company, RE Power, for € 1.34 billion. These acquisitions not only gave Suzlon an access to state of the art foreign technology for manufacturing larger (up to 5 MW) wind turbine generators but also greater access to the European market (Ramamurti & Singh 2010). Again it appears that the series of acquisitions that Suzlon undertook were part of its asset augmentation strategy. Through these acquisitions Suzlon acquired technological know-how to manufacture high capacity wind turbine generators; skilled human resources from Sudwind, Hansen, and RE Power, and market share and brand names of Hansen and RE Power.

Tata Motors

Tata Motors, a part of India's prominent conglomerate business group Tata, is India's largest automobile manufacturer. According to the company website it has sold over 7.5 million vehicles in India (Tatamotors 2012). Tata Motors is also one of the most innovative MNEs from India having made the world's cheapest car, the Tata Nano. From its inception in 1945, the company served the large commercial automobile market along with Ashok Leyland, another Indian firm set up in 1948. For many years the automobile sector in India was one of the most heavily regulated industries. As a result both firms enjoyed dominant market positions until 2002 when foreign automobile manufacturers were allowed to enter India (Bloodgood 2007). However, by this time Tata Motors had already become a large multinational enterprise. In the post liberalisation period, building on its production capacity, local brand name and technological base, Tata Motors gradually and successfully entered into the production of passenger vehicles within India. In this process besides protection in the domestic economy, Tata Motors also benefitted from its foreign linkages.

A managing director of one of the Tata group companies, Mr Harish Bhat in his book titled *Tata Log* writes that in order to distinguish its passenger cars from the contemporary cars available in India, Tata used foreign linkages in Europe. He quotes Mr Ratan Tata, who said:

> I believe totally in our own capabilities. But when we visit motor shows abroad, we see the great strides which global car companies have made, the excellent designs they have already launched in Europe and other Western countries. Our efforts should be to leapfrog into the future. For this to happen we should work in the design environment of Europe, where the design ethos is well ahead of India. Then we can hope for a car which is ahead of its time (Bhat 2012: 26).

This quote shows that Tata Motors values linkages with foreign partners who offer a range of tangible and intangible resources. Studies on the Indian automobile industry (such as Okada 2004) also argue that Indian automobile industries in general, including the small auto ancillary firms, have benefited from inter-firm linkages in the globalisation of the Indian economy. Table 11.3 summarises the benefits to Tata Motors from some prominent linkages with foreign partners.

Besides the above linkages, creation of the Tata Nano depended heavily on collaboration with the German firm Bosch, which supplied the engine, gearbox and braking system for the car. In order to work on the Tata Nano, Bosch established a research and development centre in India to collaborate with Tata Motors. Although learning from the Tata Nano was mutual, it is unlikely that the Tata Nano would have become a reality without the support of Bosch (ET 2008). Today Tata Motors is the fourth largest truck manufacturer and the second largest bus manufacturing company in the world. The portfolio of cars offered by Tata Motors range from the

Table 11.3. Tata Motors' prominent foreign linkages (compiled from www.tatamotors.com).

Foreign partner (country) – time	Type of linkage	Resources gained
Daewoo (South Korea) - 2004	Acquisition	Technology and manufacturing capacity for heavy tonnage trucks; brand and manufacturing capacity
Hispano Carrocera (Spain) - 2005	Acquisition	New commercial vehicle range
Marcopolo (Brazil) - 2006	Joint venture	Bodybuilding and design
Land Rover and Jaguar (UK) - 2008	Acquisition	Luxury car brand and technology
Trilix (Italy) - 2010	Acquisition	Design capabilities

Tata Nano to the premium vehicle brands of Land Rover and Jaguar. As a result of these assets and capabilities, Tata Motors competes with global automobile giants, such as Mercedes-Benz and Volkswagen.

Hero Group

The final case study of Hero group presents a similar story. The group started in 1944 as a cycle part trading business, however, later it also started manufacturing and assembling full cycles. In 1978, the group diversified into mopeds. Later in 1984, the group started manufacturing motorcycles with the help of Honda Motors, the well known Japanese MNE. In this case also protections provided by the state supported monopolistic access of the firm to the Indian market. This is verified from the following quote taken from 'India's New Capitalists', by Mr Harish Damodran, a famous business journalist. He notes that: 'a crucial element of their [Hero Group] success has been the state's role in guaranteeing protection against imports, and, equally important, against potential 'jettisoning' by the foreign collaborator' (Damodran 2008: 71).

The Hero group benefitted significantly from their foreign (Japanese) joint venture partner Honda, who provided them the crucial engine technological know-how for building motor bikes. However, unlike other cases where acquisitions and technology licensing was used, in this case the Hero group relied on 'equity-cum-technical tie-up' with Honda group. This is primarily because capital availability along with technological know-how, were both limitations for the Indian firm. Thus, this firm was more vulnerable than other cases. However, state protection was available in the form of a mandate which requires 'any foreign company wanting to set up shop independently to first get a no-objection from its local partner in an existing venture' (Damodaran 2008: 71).

Provisions like this enabled local Indian firms like Hero to increase their bargaining power with a foreign partner and access/learn technological know-how provided by the foreign partner while simultaneously building up their marketing networks over time to prepare themselves for global competition. The Hero Honda joint venture no longer exists: Honda exited the joint venture in December 2010. The Hero group now manages the motorcycle business under the name Hero Motocorp and has successfully mastered the business. According to recent sales figures, Hero MotoCorp achieved its highest ever monthly sales at 556,644 units in May 2012, registering a growth of 11.28% (BusinessLine 2012). Mr Anil Dua, Senior Vice President (Marketing and Sales) of Hero MotoCorp, commenting on rising sales numbers said that his company has been constantly setting up new industry benchmarks which is an indication of the company's progress and the trust that consumers have in them (BusinessLine 2012). Recently, Hero Motocorp has announced its ambitions to internationalise in Latin America with a target to acquire 10% of turnover from international business in the next five years (ET 2013).

Summary and conclusion

This chapter has examined the growth trajectory of Indian MNEs in two periods: pre-liberalisation and post liberalisation. Using international business perspective and three short cases the chapter argued that the state's protectionist policies during the pre-liberalisation period allowed Indian MNEs to grow by monopolistically servicing the large sheltered home market. In the pre-liberalisation period, the Indian economy was relatively closed. Indian policy makers were reluctant to use foreign capital for the development of economy. Imports and exports were also heavily restricted. Though these policies adversely affected efficiency and export intensity of Indian MNEs, they allowed them a monopolistic access to the large domestic home markets. As a result, many Indian MNEs earned significant financial reserves and grew in the pre-liberalisation period.

In the post-liberalisation period, when exports and imports, and inflows and outflows of foreign direct investment were liberalised, foreign MNEs entered the Indian economy. The main attraction of foreign MNEs' entry in a developing country is to exploit large unsaturated markets and to lower the cost of production. The expansion of MNEs' supply chains in a host developing country creates opportunities for local firms to link up with them by acting as suppliers, contractors or joint venture partners. Linkages with foreign MNEs are mile stones in the growth trajectory of Indian MNEs because they provide multiple opportunities for growth by learning and accessing resources from foreign MNEs.

Along with growth from linkages, Indian MNEs also acquired foreign firms. Financial resources accumulated by Indian MNEs supported their asset augmentation strategies. These strategies are reflected in the recent acquisition of foreign MNEs by Indian MNEs; for instance, acquisition of Land Rover and Jaguar which provided Tata Motors an access to advanced automobile technology for producing premium automobiles and global brands. The main aim of a developing country MNE in acquiring strategic assets is to compensate for resource deficiencies that hinder its growth. Three cases illustrated in this chapter portray different ways in which these firms benefited from strategic assets obtained by linkages and acquisitions of foreign MNEs.

The growth trajectory of Indian MNE portrayed in this chapter presents divergence and convergence with the growth trajectory of MNEs from other Asian countries. Divergence occurs due to the size of home market and convergence occurs due to underdevelopment and low cost of production. Many Asian economies, such as South Korea, Malaysia, and Japan are smaller countries; thus, they have a relatively smaller market size in comparison to home market in India. This might not have allowed MNEs in these countries to grow by servicing the home market as it happened for Indian MNEs. As a result, MNEs from small Asian countries were always outward looking; forced to grow by competing with the global MNEs. For instance, Samsung grew by competing with global MNEs, such as Sony and Apple. In contrast, Chinese MNEs have large home market and many Chinese MNEs, such as Haier has grown by servicing the home market first. Despite many divergences, the growth trajectory of most MNEs from Asian counties, such as South Korea, Taiwan, China, and India can be explained by using the asset augmentation and foreign linkage, leverage and learning strategies.

Bibliography

Agarwal, S. & Ramaswami, S.N. 1992. Choice of foreign market entry mode: Impact of ownership, location and internalization factors. *Journal of International Business Studies* 23(1): 1-27.

Akamatsu, K. 1962. A historical pattern of economic growth in developing countries. *The Developing Economies* 1: 3-25.

Athreye, S. & Kapur, S. 2009. Introduction: The internationalization of Chinese and Indian firms – trends, motivations and strategy. *Industrial and Corporate Change* 18(2): 209-209.

Athukorala, P.C. 2009. Outward Foreign Direct Investment from India. *Asian Development Review* 26(2): 26-56.

Balasubramanyam, V.N. & Mahambare, V. 2003. Foreign direct investment in India. *Transnational Corporation* 12(2): 45-72.

Barkema, H.G., Bell, J.H.J. & Pennings, J.M. 1996. Foreign Entry, Cultural Barriers, and Learning. *Strategic Management Journal* 17(2): 151-166.

Bhat, H. 2012. *Tata Log.* Penguin, New Delhi, India.

Bloodgood, L. 2007. *Competitive conditions for foreign direct investment in India.* Federal Publications, Washington, DC, USA.

Buckley, P., Enderwick, P., Forsans, N. & Munjal, S. 2012. *Internationalisation of the Emerging Country Multinational Enterprise: Role of Foreign Resources and Learning.* The Great Diversity – Trajectories of Asian Development, World University Network (Leeds-Bergen) Conference. Bergen, Norway.

Buckley, P.J. & Casson, M. 1976. *The future of the multinational enterprise.* Macmillan, London, UK.

Buckley, P.J., Forsans, N. & Munjal, S. 2009. *Foreign acquisitions by Indian multinational enterprises: a test of the eclectic paradigm*. Cambridge Symposium on International Manufacturing, Møller Centre, Cambridge, UK.

BusinessLine 2012. *Hero MotoCorp sales in May spurts 11%*. Business Line. The Hindu, New Delhi, India.

Damodaran, H. 2008. *India's New Capitalists*. Palgrave Macmillan, Hampshire, UK.

Desai, A.V. 1972. Technology management in Indian companies. *Long Range Planning* 5(4): 70-72.

Dunning, J.H., Kim, C. & Park, D. 2008. Old wine in new bottles: A comparison of emerging-markets TNCs today and developed country TNCs thirty years ago. In: Sauvant, K.P. (ed.). *The rise of transnational corporations from emerging markets: Threat or opportunity?* Edward Elgar, Cheltenham, UK: 158-180.

The Economic Times. 2002. Pre-liberalisation cos dominate in India. Bennett, Coleman & Co. Ltd. Available at: http://articles.economictimes.indiatimes.com/2002-05-21/news/27365686_1_software-companies-marine-products-computer-software.

The Economic Times. 12 January 2008. What forced Bosch to go Nano? Bennett, Coleman & Co. Ltd. Available at: http://articles.economictimes.indiatimes.com/2008-01-12/news/28414252_1_bernd-bohr-bosch-crdi.

The Economic Times. 30 January 2013. Hero MotoCorp goes global with operations in 3 Latin American countries. Bennett, Coleman & Co. Ltd. Available at: http://articles.economictimes.indiatimes.com/2013-01-30/news/36636083_1_hero-motocorp-brand-hero-global-brand.

Gopinath, S. 2007. *Overseas Investment by Indian Companies: Evolution of Policy and Trends*. Reserve Bank of India, Mumbai, India.

Greenaway, D., Morgan, W. & Wright, P. 2002. Trade liberalisation and growth in developing countries. *Journal of Development economics* 67(1): 229-244.

Khanna, T. & Palepu, K. 2010. *Winning in Emerging Markets*. Harvard Business Press, Boston, MA, USA.

India Knowledge@Wharton. 14 December 2006. *How Indian Companies Fund Their Overseas Acquisitions*. Available at: http://knowledge.wharton.upenn.edu/india/article.cfm?articleid=4131.

Kotwal, A., Ramaswami B. & Wadhwa, W. 2011. Economic liberalization and Indian economic growth: What's the evidence? *Journal of Economic Literature* 49(4): 1152-1199.

KPMG 2012. Emerging Market International Acquisition Tracker. Available at: http://www.kpmg.com/Global/en/IssuesAndInsights/ArticlesPublications/high-growth-markets-tracker/Documents/hgm-tracker-august-2012-data.pdf.

Luo, Y. & Tung, R.L. 2007. International expansion of emerging market enterprises: A springboard perspective. *Journal of International Business Studies* 38(4): 481-498.

Makino, S., Lau, C.-M. & Yeh, R.-S. 2002. Asset-Exploitation versus Asset-Seeking: Implications for Location Choice of Foreign Direct Investment from Newly Industrialized Economies. *Journal of International Business Studies* 33(3): 403-421.

Mathews, J.A. 2002. *Dragon multinational: A new model for global growth*. Oxford University Press, USA.

Mathews, J.A. 2006. Dragon multinationals: New players in 21st century globalization. *Asia Pacific Journal of Management* 23(1): 5-27.

Narula, R. & Nguyen, Q.T.K. 2011. *Emerging country MNEs and the role of home countries: separating fact from irrational expectations*. UNU-MERIT Working Paper Series 021, United Nations University, Maastricht Economic and social Research and training centre on Innovation and Technology.

Okada, A. 2004. Skills Development and Interfirm Learning Linkages under Globalization: Lessons from the Indian Automobile Industry. *World Development* 32(7): 1265-1288.

Pfeffer, J. & Nowak, P. 1976. Joint ventures and interorganizational interdependence. *Administrative Science Quarterly* 21(3): 398-418.

Porter, M.E. 1986. *Competition in global industries*. Harvard Business Press, Boston, MA, USA.

Pradhan, J.P. 2007. Growth of Indian multinationals in the world economy: implications for development. *ISID Working Paper No. 2007/04*. Available at: SSRN: http://ssrn.com/abstract=1515650 or http://dx.doi.org/10.2139/ssrn.1515650.

Ramamurti, R. & Singh, J.V. 2010. *Emerging Multinationals in Emerging Markets*. Cambridge University Press, Cambridge, UK.

Reserve Bank of India. 2012. Outward Indian FDI – Recent Trends and Emerging Issues. Speech delivered by Shri. Harun R. Khan, Deputy Governor, Reserve Bank of India at the Bombay Chamber of Commerce & Industry, Mumba, March 2, 2012. Available at: http://rbi.org.in/scripts/BS_SpeechesView.aspx?Id=674.

Stucchi, T. 2012. Emerging market firms' acquisitions in advanced markets: Matching strategy with resource-, institution- and industry-based antecedents. *European Management Journal* 30(3): 278-289.

Suzlon 2012. Company Profile. Available at: http://www.suzlon.com/about_suzlon/l2.aspx?l1=1&l2=1.

Tatamotors 2012. Company Website. Available at: www.tatamotors.com.

Chapter 12: Renewable energy and emerging low carbon development strategies in East Asia

Christopher M. Dent

The twin challenges of climate change and energy security are compelling East Asian states to develop stronger renewable energy and other green energy sectors as part of their low carbon development strategies. Energy has always played a fundamental role in economic development, and the cleaner and more sustainable nature of renewable energy (RE) systems makes them crucial to securing low carbon futures. The RE sector has expanded significantly worldwide over the last few years, and in East Asia faster than any other region. RE has furthermore become one of the defining features of East Asia's new industrial policy and 'new developmentalism', founded more generally on new configured forms of state capacity shaped in response to various challenges confronting the region's nations. Studying the recent progress of East Asia's RE sector provides useful insights into these key developments in East Asia's political economy, and the region's prospects for transition towards low carbon development: a particularly difficult challenge given the many high carbon-intensive aspects of the East Asia's economy.

The emergence of East Asia's 'new developmentalism'

The ascendance of East Asia's renewable energy sectors is linked to a broader 'new developmentalism' to have emerged in the region. Developmentalism is a much applied concept to explain the state's role in East Asia's contemporary economic development, and more specifically state-business and state-society relations in various development processes. Its main theoretical roots lay in the concepts of the developmental state, state capacity and socialist market thinking (Amsden 1989, Johnson 1982, Rodan 1989, Suliman 1998, Weiss 1998, Woo-Cumings 1999). Taking the first and arguably most important of these concepts, developmental states are fundamentally capitalist economies with three general attributes: (1) state institutions with the capacity to strategise and plan a transformative economic development path, with institutional and planning co-ordination led by a 'pilot agency'; (2) strategic policies to operationalize these plans and realise their transformative economic goals that change over time in accordance with developmental progress; and (3) developmental partnerships forged by the state primarily with the business sector but also with society, whereby non-state actors (mostly firms) are co-opted into transformative economic projects through various institutionalised consultative

C. Brautaset and C.M. Dent (eds.) **THE GREAT DIVERSITY**
DOI 10.3920/978-90-8686-786-8_12, © Wageningen Academic Publishers 2014

mechanisms. Similarly, in socialist market economies like China and Vietnam, the state also possesses strategic economic planning institutions and policies but where the state plays a substantial interventionist role in the economy through more direct control of markets and the means of production, yet still harnessing the power of private enterprise to help realise strategic economic objectives albeit through less consultative means than in developmental states. The state's capacity to bring about prosperity-generating transformative change is a core issue that links both concepts.

The rationale for state support for renewables is based on public good and positive externality arguments. For example, the expansion of renewable energy practices will reduce carbon emissions, lead to cleaner air, and mitigate society's supply risk dependency on exhaustible fossil fuel resources. These state actions constitute the provision of public goods where the market (i.e. private enterprise) alone is unable to sufficiently deliver these welfare-enhancing outcomes, and more broadly placing the economy and society on track to a low carbon future. Yet it has only been in recent years that East Asian states have sought to substantively support renewable energy development. Moreover, from the 1980s onwards the effects of globalisation, ascendant neo-liberalism and socio-technological change caused a retreat of East Asian developmentalism. The adoption of neo-liberal policies in East Asia especially led to less proactive state economic management and a corresponding greater faith placed in market mechanisms (Pirie 2008, Radice 2008).

Yet as the increasing risks associated with globalisation and neo-liberalism (e.g. financial crises principally caused by deregulation, commodity price volatility) became more evident during the 2000s, so East Asian states began to revert back to developmentalist solutions and reinvigorated exercises of state capacity (Hayashi 2010, Gills 2010, Karagiannis & Madjd-Sadjadi 2007, Stubbs 2009, Weiss 2010). Furthermore, the emergent global challenges of climate change and acute energy security predicaments in the region has led many East Asian states to place renewable energy and green energy technologies more generally at the centre of a 'new developmentalism' founded on the new transformative goal of low or lower carbon development. This has entailed the incorporation of wind, solar, bio-energy, geothermal, hydro, ocean and other RE sectors more deeply into the new industrial policies of East Asian states, not just for environmental reasons but also because of strategic economic motives, these being industries with huge potential for expansion in the 21st century. The tensions between these two drivers – the environmental and the strategic economic – is a key theme of this paper. Indeed, viewing renewable energy sectors as 'industries' may appear somewhat oxymoronic due to the strong connotations that 'industry' has with energy-intensive and pollutive forms of material-based productive activity. Certainly, most mainstream industries (e.g.

textiles, steel and automobiles) remain firmly associated with high-carbon energy industrialisation and economic 'modernisation' *per se*, whereby nature's power and resources are subjugated to the desire for material outputs that 'satisfy' human demands. Thus, the extent to which East Asia's 'new developmentalism' represents a new post-modernist alignment of economic development and ecological balance is a key subject for debate.

Developments in renewable energy globally and in East Asia

Renewable energies are derived from replenishable natural processes, sources or phenomena, such as wind, solar, geothermal, hydropower, ocean, biomass and biofuels. Their ability to produce clean energy – whether for electricity generation, thermal heat, or motive power – makes renewables an essential element of low carbon development and climate change strategies. For nations depending heavily on imported energy fuels (e.g. Japan, South Korea), renewables have the added advantage of being inherently indigenous energy sources, thus helping mitigate foreign supply risks in addition to environmental risks.

Between 2005 to 2012, worldwide renewable energy power capacity increased from 931 gigawatts (GW) to 1,472 GW and from 15.6% of total world electricity generation to 21.7% (Table 12.1). In terms of global primary energy consumption (electricity, motive power, thermal heat), renewables accounted for 19.0% of the total in 2011, with traditional biomass being the most prominent RE sector (9.3% of the global total) while nuclear power accounted for 2.8% and fossil fuels 78.2%. From 2008 to 2012, renewables also accounted for half of all investment in new energy generation capacity worldwide (REN21 2013).

Table 12.1 shows that wind and solar photovoltaic (PV) have been the fastest growing and now most significant new RE sectors, although the long established hydroelectric sector remains by far the most dominant, being responsible for two-thirds of the renewables total worldwide. Perhaps most significantly, wind and solar energy are now scaled-up and fully commercialised industries with the combined potential to provide more off-grid and grid-connected electricity than any other RE sector in the decades to come. By 2012, the combined installed electricity capacities of solar and wind overtook nuclear power for the first time (Table 12.1). The pace of techno-innovatory advances made in these two renewable energy sectors have been staggering (Lee *et al.* 2009), and in some countries they are competing on parity terms with fossil fuels in power generation (REN21 2013). They also have cluster linkages to various other high-tech sectors. In wind energy's case, this includes nanotechnology (composite materials), aerospace (aerodynamics and advanced

Table 12.1 Renewable energy sector development, World and East Asia overview (Energy Information Administration (2010), EPIA (2012), GWEC (2012), IEA (2012), REEEP (2012), REN21 (2006, 2010, 2013)).

Sector	Electricity generation (GW installed capacity, world)			Main East Asian producers, 2012 (installed capacity level, world ranking)
	2005	2012	Added	
Wind	59.0	282.6	223.6	China 75.3GW (1) Japan 2.6GW (13) Taiwan 0.6GW (28)
Solar Photovoltaic	5.4	102.1	96.7	China 8.3GW (3) Japan 6.9GW (5) South Korea 1.0GW (14)
Concentrating Solar Power	0.4	2.5	2.1	Thailand 5MW (8)
Hydroelectric	816.0	990.0	174.0	China 229.0GW (1) Japan 27.2GW (9)
Geothermal	9.3	11.7	2.4	Philippines 1.9GW (2) Indonesia 1.3GW (3) Japan 0.5GW (8)
Biomass	44.0	83.0	39.0	China 8.0GW (4) Japan 3.3GW (5) Thailand 1.3GW (19)
Wave	0.0	0.0	0.0	Sector still in experimental stage Japan as one of the world's technological leaders
Tidal	0.3	0.5	0.2	South Korea 0.25GW (1)
Renewables sector total	931 (15.6% total)	1,472 (21.7% total)	389	China 319GW (1) Japan 39GW (8)
Nuclear sector	377	381 [a]	4	
Fossil fuel sector	4,659	4,921	262	
All sectors	5,967	6,774		
Bioethanol	33bn litres	83bn litres		China 2.1bn (3) Thailand 0.7bn (7)
Biodiesel	4bn litres	23bn litres		Indonesia 1.5bn (6) Thailand 0.9bn (7)

[a] 2010 figure for nuclear.

materials), energy storage electronics (advanced batteries) and meteorology software. Solar energy meanwhile has close links with advanced materials and chemicals, satellite guidance technology, electronics, lasers, and high-grade concentrators used in space technology applications. Bioenergy (comprising biomass and biofuels) are another important sector for both electricity generation as well as transportation and thermal energy. Many renewables remain, though, relatively small scale. For instance, global installed capacity for geothermal was only 11.7 GW in 2012, around half that of China's Three Gorges Dam plant. Ocean energy (wave and tidal) is still a largely experimental micro-sector with a global power capacity of just 0.5 GW in 2012, as is concentrating solar power (CSP) with only 2.5 GW total installed capacity although this is a very fast growing sector.

There are, however, a number of fundamental constraints and contentious issues associated with renewable energies. First, the intermittent power problem means for example that on-shore wind turbines normally operate at 20 to 40% of their technical capacity, solar power only produces energy during daytime hours, and ocean power is dependent on wind speeds, currents and tidal movements. A second related problem is the non-dispatchable nature of RE sectors like wind, solar and ocean power where energy output cannot be stored but normally only used at immediate time of generation. Hydroelectric, geothermal and biomass are able to circumvent this problem through tapping into constant energy streams, thus having more predicable load factors. Due to intermittency and non-dispatchable issues, wind, solar and ocean energy are likely to remain as supplements to more consistent and predictable forms of energy generation. A third problem concerns various environmental and sustainability predicaments facing all renewables, such as acute ecological problems caused by large hydroelectric dams, emissions from biomass and biofuel combustion, visibility and noise nuisances related to wind farms, spatial scale constraints (especially for wind and solar farms), and the dependency of certain sectors on finite natural materials, e.g. lithium for solar panels, rare earth minerals for wind turbine magnets. Although such constraints and issues are becoming increasingly apparent as renewables are scaled up, rapid technological advances in many RE sectors are simultaneously providing solutions to them. For instance, solar PV panel prices are currently a fraction of what they were just a few years ago and their power generation costs have also fallen drastically.

Most renewable energy sectors have grown faster in East Asia than anywhere else worldwide, and China deserves special attention in this regard. By 2012 it had developed the world's largest capacity for renewable electric power at 319 GW, the United States coming a distant second at 164 GW and Japan eighth with 39 GW (REN21 2013). Over recent years China has been responsible for around 40% of new additions to global renewable energy capacity. In 2010, China overtook the US as having the world's largest installed capacity for wind energy; in 2012 China's wind energy output overtook that of its nuclear sector and added more electricity capacity to the nation's grid than coal. Meanwhile Japan is positioned fifth worldwide on electricity generation from both solar PV and biomass, and is a top ten producer for hydroelectric and geothermal. The Philippines and Indonesia hold second and third global positions respectively behind the US on geothermal, while South Korea has emerged as a world leader in tidal energy as well as an emerging player in solar PV. Furthermore, there has been a marked shift in the manufacture of RE equipment to East Asia, and particularly to China which by the early 2010s was producing two-thirds of the world's solar PV cells and modules, a third of all wind turbines and around four-fifths of solar hot water collectors. Japanese companies remained technological leaders in wind, solar PV and bioenergy sectors.

Renewable energy policy in East Asia

By early 2013 there were reported to be 127 countries pursuing a renewable energy policy and those from East Asia have been amongst the most active (REN21 2013). While RE policy in East Asia is mostly focused on business and industry – and hence its primary classification as industrial policy – it can extend to society generally by involving households and communities as both grid-connected and off-grid renewable energy producers. Japan was the first country from the region to initiate a modern RE policy (i.e. beyond large hydroelectric) from the early 1970s onwards, mainly in response to the 1973/74 oil crisis. Other nations from the region followed later that decade and into the early 1980s: the Philippines in 1978, Singapore (1979), Malaysia (1981), and South Korea (1985). China, Indonesia and Thailand first developed their RE policies in the 1990s, and Vietnam from 2001 (IEA 2012, REEEP 2012). RE policy instruments can be categorised as follows:

- *Regulatory mandates* establish legally binding requirements on firms to undertake particular actions. For example, renewable portfolio standards (RPS, which have been gradually introduced into East Asia) and ad hoc mandatory obligations oblige energy producers to source certain amounts of their electricity generation from renewable energies, and mandatory obligations may also work on a similar principle in the construction and biofuels industries, or some other stipulation.
- *Direct financial support* instruments comprise state subsidies, grants, loans, and capital investment in RE sector plants and infrastructure. Subsidies and grants have remained a particular important feature of RE policy for most East Asian nations.
- *Market-based instruments* adapt or use the market mechanism to provide a variety of different financial incentive measures. These include competitive bidding,

tradable permits or certificates and various kinds of tax incentives. Feed-in tariff (FIT) systems have proved increasingly popular, whereby the government pays premium tariff rates to normally small-scale suppliers of renewable energy under long-term contracts.

The above policy instruments also form an essential part of increasingly substantive and coherent *strategic planning* on RE sector development in East Asia. In the industrial policy context, governments have over time promoted particular renewables over others or adopted a broad multi-sector approach, where renewables in general have been selected for promotion. Development targets are normally set in terms of actual installed power capacity, percentage share of total energy production, or market size or share. While East Asian nations do not possess the world's highest-level targets on renewables' share of the total energy mix, their level of strategic ambition is very high when compared to similar income-level countries worldwide (REN21 2013).

South Korea and Taiwan have set the broadest range of RE sector-specific targets, followed by China, Indonesia, the Philippines and Thailand (Tables 2 and 3). All nations (except Singapore) have set such targets for at least one or two specific sectors as well as an overall RE sector target in terms of power capacity or share of the energy mix. Japan and South Korea failed to meet more or less all its past targets during the 2000s. While China well exceeded its wind and solar energy targets over the same time, it failed to realise its general 11th Five-Year-Plan (FYP) non-fossil energy target of 10% of primary energy consumption by 2010, achieving only 8.4%, and this including nuclear.

Up to the mid-2000s, RE policy in East Asia had developed in a somewhat piecemeal and ad hoc manner but greater coherence came after governments in the region introduced their strategic master plans specifically for renewables, namely:

- China's Medium and Long-Term Development Plan for Renewable Energy to 2020
- Japan's Cool Earth Energy Innovative Technology Plan to 2050
- Malaysia's Renewable Energy Development – 9th Malaysia Plan 2006-2010, and thereafter the National Renewable Energy Policy and Action Plan – 10th Malaysia Plan 2011-2015
- Philippines' Renewable Energy Policy Framework 2003-2013
- South Korea's New Renewable Energy Medium-Term Plan 2010-2015
- Taiwan's New Energy Policy to 2025
- Thailand's Alternative Energy Development Plan 2012-2021
- Vietnam's Renewable Energy Master Plan 2011-2015 – 9th Five-Year Plan 2011-2015

This corresponded to the more programmatic integration of strategic planning on renewables into broader macro-development plans and consequently becoming a key feature of the East Asia's new developmentalism. Renewables have too become increasingly salient in East Asia's national energy strategy plans[151] and national climate change plans.[152] This marks a growing conflation regarding strategic planning on energy security, climate change, industrial policies and macro-development, with renewable energy being a pivotal link between all these policy domains.

Concerning the *main technical emphasis* of RE policy generally, developed nations (Japan, South Korea, Singapore) have fostered a more *technology-oriented* approach where the main emphasis has been on research, development and dissemination (RD&D) support to improve firms' technological and innovatory competitiveness, and the technical efficacies of RE installations. This is consistent with the technology cluster approach – consistent with past developmental state practice – where the state has promoted the development of core 'new energy' technologies to be diffused through multiple industrial sectors, such as Japan and South Korea's investments in fuel cells for the automotive and ICT industries. The development of energy efficiency and saving technologies has also been strongly prioritised by the countries. In sum, this may be considered as much an innovation policy as an industrial policy (Noland 2007). The region's developing nations (China and most of Southeast Asia) have, though, adopted a more *installed capacity-oriented* RE policy where realising production or power generation based goals are the priority (Table 12.2). These countries generally lack new technology development capacity, although China, Malaysia and Thailand are increasingly looking to strengthen their RE sector innovative capabilities.

[151] As evident in: Japan's Strategic Energy Plan, first formulated in 2003 and thereafter revised on a 3-year cycle, the last being in 2010, with policy co-ordination links with the later discussed 2010 New Growth Strategy for the Japanese economy (ARNE 2010); South Korea's National Energy Plan 2008-2030 is integrated into the also later discussed Green Growth macro-development programme; Singapore's Comprehensive Blueprint for Clean Energy (2007); Indonesia's National Energy Blueprint 2005-2025; Vietnam's National Energy Development Strategy 2008-2020; China's national energy plans, as integrated into its 5-Year Plan framework.

[152] For example, China's climate change strategy, introduced in 2007, has close policy co-ordination links with the country's Medium and Long-Term Development Plan for Renewable Energy established in the same year (NRDC 2007a, 2007b).

Table 12.2. Renewable energy policy in East Asia – overview (IEA (2012), Martinot and Li (2010), Olz and Beerepoot (2010), REEEP (2012), REN21 (2013), author's research).

	China	Indonesia	Japan	Malaysia	Philippines	Singapore	South Korea	Thailand	Vietnam
General policy instruments									
Regulatory mandates									
Renewable portfolio standards (RPS)	•		•		•		•		
Mandatory obligations (e.g. prod. level)	•	•			•	•			
Direct financial support									
Subsidies, grants, loans	•	•	•	•	•	•	•	•	•
Capital investment (infrastructure, etc.)	•	•	•	•			•		•
Market based									
Competitive bidding	•				•	•			
Tradeable permits or certificates		•	•		•				
Feed-in tariffs (FiT)	•	·	·	•	•		•	•	
Tax incentives	•	•	·	•	•	•	•	•	•
Strategic Planning									
Target setting									
Solar	•	•	•		·		•	•	·
Wind	•	•			•		•	•	·
Hydro (large, or general)	•			·	•			•	
Hydro (small)		•							•
Geothermal		•			•				
Biomass	•	•			·			•	·
Biofuels	•	•			•				•
Ocean (wave, tidal)					·				
Waste-to-energy	•								
General All Sectors Target	•	•	•	•	•		•	•	•
Strategic development plans									
Renewable Energy Master Plan (all sectors)	•	•		•	•		•	•	•
Sector-Specific Renewable Energy Plans	•	•	•	·	•	·	•	·	•
Main technical emphasis									
Technology-Oriented - strong emphasis on R&D			•			•	•		
Installed Capacity-Oriented - strong emphasis on exanding installed RE generation capacity	•	•		•	•			•	•

Small dots indicate relatively minor significance of the factor in question, such as an FIT system that only applies to one RE sector (e.g. solar PV in Japan) but not others.

Renewable energy and East Asia's new developmentalism: three case studies

Japan

Japan was East Asia's pioneer developmental state (Johnson 1982). During the 1990s, Japan's industrial policies became less focused on specific sectors and more on macro-structural reform and developing generic technology clusters, this accordingly shaping the technology-oriented nature of Japan's RE policy (METI 2010a). Into the 2000s, Japan pursued in METI's own words a 'productivity-oriented economic policy that was based on excessive market fundamentalism and was overly tilted toward the supply side' (METI 2010b: 2), where industrial policy both lacked substance and had limited impact, and in which no coherent or lasting macro-development strategy was devised. However, most recently a new developmental approach was introduced in 2010 by the Japanese government's New Growth Strategy (NGS) that 'aims to achieve economic growth by turning the problems faced by the economy and society into opportunities for creating new demand and employment' (*ibid*: 2).

A core element of the NGS strategy is to strengthen Japan's industrial-technological capabilities in seven strategic areas up to 2030, environment and energy ('green innovation') being the first listed, the other six being medical/health care, finance, science and technology nation, tourism/local revitalisation, human resources, and Asian regional economic integration. These are underpinned by 21 national strategic projects, one of which under 'green innovation' is the expansion of renewables sector into a JPY 10 trillion (US$ 110 billion) market by 2020. The NGS is augmented by the Industrial Structure Vision that sets broad aspirational objectives for diversifying Japan's industrial base towards a multi-sector low carbon economy that includes a relatively prominent role for renewables (METI 2010c). It also incorporates the installed capacity target on solar PV (14 GW by 2020, 53 GW by 2030) from the aforementioned Cool Earth Energy Innovative Technology Plan.

The main goal of the NGS is the 'creation of a low-carbon society through a comprehensive policy package including new systems design, systems changes, new regulations, and regulatory reform, and to support the rapid spread and expansion of environmental technologies and products' (METI 2010b: 20). This transformation would include 'measures to support the spread and expansion of renewable energies (solar, wind, small-scale hydroelectric, biomass, geothermal, etc.) by expanding the electric power feed-in tariff system' (*ibid*: 21-22). More generally, NGS employs other policy methods and measures such as public-private partnerships and consultation mechanisms, promoting corporate and industrial clusters in strategic technology areas, establishing Japanese industrial and technology standards as international ones in 'energy management', and the construction of a new research system based on the industrial-academic-government research complexes. The NGS' overarching low carbon development goal was to reduce worldwide greenhouse gas emissions by at least 1.3 billion tons of CO_2 equivalent by 2030 using Japanese technology: the global scope target derives from the technological contributions made by Japanese companies worldwide, and reflects the aforementioned technological and innovatory leadership of Japanese firms in many renewable and other green energy sectors. It could be argued that renewable energy is overshadowed in the NGS to some degree by other green energy industries, especially alternative energy vehicles and fuel cells. The nuclear power sector is additionally afforded high priority under the environment and energy strategic area.

China

Although less *dirigste* than before, China's FYPs are still devised largely through a complex intra-state process where private and civil sector agencies have limited influence. Managing China's increasingly acute energy security challenges have become a central theme of national strategic planning. Concurrently, fostering sustainable energy systems and economic development have too been afforded greater priority, as first substantively operationalised in the 11th FYP (2006-2010) and continued in the current 12th FYP (2011-2015), its other overarching aims being to achieve more equitable income growth, promote domestic consumption, improve social infrastructures and safety nets, and strengthen China's innovatory capabilities (NDRC 2011). Generally speaking, there is growing acknowledgement in China of the imperative to find low carbon solutions to the country's many development challenges (CCICED 2010, Wang & Watson 2009).

The 11th FYP incorporated China's Renewable Energy Development Targets set at the start of the plan period in 2006 of increasing RE's contribution of primary energy consumption to 15% by 2020 (up from 7.5% achieved by 2005), with various sector-specific targets set, for example creating 30 new large-scale wind farms by 2010. Meanwhile, the Medium and Long-Term Development Plan for Renewable Energy launched in 2007 committed US$ 263 billion of public investment in RE sectors up to 2020 while simultaneously making numerous references to harnessing the market in achieving its goals (NDRC 2007a). Later, the 12th FYP (2011-2015) included a RMB4 trillion (US$ 610 billion) funded programme to promote seven Strategic Emerging Industries (SEIs) for 'clean' development and a new industry base, namely: new generation ICT, energy-saving and environment protection, new energy, biotechnology, high-end equipment, new materials and alternative energy cars. These are intended to form the new future backbone of the Chinese economy with the aim of increasing their Gross Domestic Product (GDP) share from the current 5% to 8% by 2015 and 15% by 2020 (NDRC 2011). Renewables are specified as a key component of the 'new energy' SEI sector that in addition includes nuclear power (40 GW additional capacity by 2015) and large hydro (50% increase in installed capacity by 2015). Under the SEI programme, China plans to enhance its new technological and innovation capabilities in wind energy, solar PV, fuel cells and electric vehicles as part of a broader strategy of moving China from 'world factory' to a 'innovative hub' economy. Notwithstanding the persistent fixation with expanding installed generation capacity regarding renewables, there is also a clear indication in the current FYP that China intends also to develop over time a more technology-oriented RE policy. The 12th FYP also set China's current sector-specific targets on renewables (Table 12.3).

South Korea

South Korea was one of East Asia's most ardent developmental states until it dismantled its FYP framework in 1993 and thereafter shifted towards a more neo-liberal economic ideology and policy approach which persisted throughout most of the 2000s (Lim 2010). However, a return to developmentalism came with the introduction of the 'Green Growth Strategy' (GGS) macro-development programme in 2008 under the Lee Myun-bak administration. The GGS is a medium and long-term strategic plan to create a low carbon society and green growth economy (Jung & Ahn 2010, KEITI 2009). Its primary transformative goal is to make South Korea the world's 'seventh green power' by 2020, and the 'fifth green power' by 2050, to be achieved through new FYPs co-ordinated by the newly formed Presidential Committee on Green Growth, comprising governmental, business and civil societal representatives. The plans are based on three main strategic approaches and 10 strategic projects implemented across them. Renewables are included under the first strategic approach (*measures for climate change and securing energy independence*) and first strategic project (*reducing fossil fuel use*), where specific targets are set of increasing renewable and new energy supply use from 2.7% in 2009 to 3.8% by 2013, and then a target extending beyond the five-year period of 6.1% by 2020, and 11.5% by 2030 (Kwon 2010, Moon 2010).

GGS targets and other objectives on renewable energy are co-ordinated with South Korea's National Energy Plan 2008-2030. Renewable energy industries are also noted under the second strategic approach (*creation of new engines for economic growth*) and the associated strategic project of developing green technologies, specific reference made here to solar, biomass and fuel cell sub-sectors from the 27

Table 12.3. East Asia's renewable energy and carbon emission targets (Energy Information Administration (2010), EPIA (2012), GWEC (2012), IEA (2012), Olz and Beerepoot (2010), REEEP (2012), REN21 (2006, 2010, 2011, 2012).

Government	General renewables	Specific targets	Carbon emission reduction targets
Northeast Asia			
China	Non-fossil fuel, primary energy (inc. nuclear): 11.4% (2015), 15% (2020); Non-fossil fuel (exc. hydro) electricity: 1% (2010), 3% (2020)	Hydro: 284 GW (2015); 300 GW (2020) Wind: 100 GW on-grid, inc. 5 GW offshore (2015); 150 GW, inc. 30 GW offshore (2020) Solar: 15 GW, inc. 1GW CSP (2015); 20 GW (2020 - unofficial) Biomass: 30 GW (2020) Biofuels: 5 million tonnes bioethanol used (2011-2015)	Energy consumption per GDP unit: -16% (2011-2015); CO_2 emission per GDP unit: -17% (2011-2015), -40 to -45% (2011-2020)
Japan	Primary energy: 10% (2020); Non-hydro electricity: 1.63% (2014), inc. hydro electricity: 10% (2014)	Solar: 4.8 GW (2010), 14 GW, 5.3m homes (2020), 53 GW (2030)	15% reduction by 2020 from 2005 emission levels
Mongolia	Electricity: 20-25% (2020)		
South Korea	General energy: 13,016 GWh, 2.9% (2015); 21,977 GWh, 4.7%, (2020); 39,517 GWh, 7.7% (2030)	Hydro: large 3,860 GWh (2030); small 1,926 GWh (2030) Wind: 16,619 GWh (2030), 2 GW offshore (2020) Solar: PV: 1,971 GWh (2030); thermal 2,046 GWh (2030) Biomass: forest 2,628 GWh (2030) Biofuels: biogas 161 GWh (2030) Geothermal: 2,803 GWh (2030) Ocean: 6,159 GWh (2030)	30% reduction compared to 'business as usual' scenario by 2020

Table 12.3. Continued.

Government	General renewables	Specific targets	Carbon emission reduction targets
Northeast Asia (continued)			
Taiwan	Electricity: 9,952 MW, 14.8% total (2025), new: 6,600 MW (2025); 12,502 MW, 16.1% total (2030)	Hydro: 2,502 MW (2030) Wind: onshore 1.2 GW (2030); offshore 3 GW (2030) Solar: 3,100 MW (2030) Biomass: 1,400 MW (2030) Geothermal: 200 MW (2030) Ocean: 600 MW (2030) Fuel cells: 500 MW (2030)	Return to 2005 emission level (2020), return to 2000 level (2025)
Southeast Asia			
Indonesia	Primary energy: 25% (2025); Electricity: 15% (2025)	Hydro: 1.3 GW new (2015), 2 GW (2025) Wind: 300 MW (2014) Solar: 2 GW (2014) Biomass: 400 MW new (2015), 810 MW (2025) Biofuels: transportation 5% (2025); biodiesel 20% diesel (2025); bioethanol 15% gasoline (2025) Geothermal: 12.6 GW (2025)	26% from business-as-usual level by 2020 based on domestic funding, or 41% by 2020 based on international funding (US$18.2 billion)
Malaysia	Primary energy: 11% (2020), 14% (2030), 36% (2050)	Solar: 1,250 MW (2020) Biomass: 1,065 MW (2020)	
Philippines	Electricity: 15,236.3 MW (2030, triple 2010 capacity level)	Hydro: 8,729.1 MW (2030) Wind: 2,378 MW (2030) Solar: 285 MW (2030) Biomass: 306.7 MW (2030) Geothermal: 3,467 MW (2030) Ocean: 70.5 MW (2030)	

Table 12.3. Continued.

Government	General renewables	Specific targets	Carbon emission reduction targets
Southeast Asia (continued)			
Singapore			16% from projected 2020 'business-as-usual' level
Thailand	Primary energy: 20.3% (2022), 14.1% electricity (2022)	Hydro: 1,608 MW (2022) Wind: 1.2 GW(2022) Solar: PV: 2 GW (2022); thermal: 100 ktoe (2022) Biomass: 3,630 MW (2022); thermal: 8,200 ktoe (2022) Biofuels: biogas: 600 MW; bioethanol: 9m ltrs/day (2022); biodiesel: 6m ltrs/day (2022)	30% from 2009 level by 2020, 42 million tonnes CO_2 equivalent in 2020
Vietnam	Primary energy: 5% (2020), 8% (2025), 11% (2050); Electricity: 5% (2020), 241MW p/year average new (2006-2015), 160MW p/year (2016-2025), 4,050MW new (2025)	Biofuels: various for bioethanol and biodiesel (2015, 2020)	

core green technologies listed overall, as well as strengthening South Korea's new energy RD&D capacity over the plan period. The core strategies and transformative goals of the GGS provide the basis for the Ministry of Knowledge Economy's (MKE) new industrial policy, where 17 strategic sectors are identified for development, renewable energy placed under the 'green technology' cluster category. The third and last strategic approach (*improving life quality and Korea's international status*) envisions South Korea to become an international leader on green growth and low carbon development. The country's environmental groups have, though, criticised the programme for lacking ambitious objectives on renewable energy development and including nuclear energy under 'green growth'.[153] Moon (2010) additionally is critical of its primary focus on economic growth and subsequent neglect of green society.

Between old and new developmentalism?

While it is evident that the goal of low or lower carbon development forms a core aim of East Asia's new developmentalism, it also vies with other strategic imperatives such as strengthening national economic competitiveness. Furthermore, many macro-development plans contain programmes or promote industries that are inherently high-carbon based and still rooted in the modernist 'old developmentalism' of the past. For example, the expansion of physical infrastructures, export-oriented manufacturing, and an emphasis on state direct financial support remain defining characteristics of China's 12th FYP, while its promotion of higher levels of domestic consumption somewhat contradict the plan's low carbon objectives. It should too be noted that China's energy policy is expanding on all fronts at relatively similar paces. For example, over the 12th FYP period (2011-2015) the Chinese government expects the number of coal-fired power stations to increase by 33%, a growth in capacity equivalent to the total current EU capacity in this sector. At the same time, the Plan aims to reduce the nation's electricity generation dependence on coal from the present 70 to 62%, meaning a comparative faster expansion of other sectors such as wind energy. Elsewhere in the region, conventional energy and chemicals are amongst the 20 strategic areas that comprise Singapore's latest industrial master-plan, which includes the further expansion of the city-state's already huge petrochemical complex and the construction of a large new liquefied natural gas terminal. In South Korea, the GGS must simultaneously contend with MKE industrial policy for upgrading the traditional 'flagship industries', including the energy-intensive sectors of shipbuilding and steel.[154]

[153] *Korea Times*, 20 August 2008; *Korea Times*, 22 August 2008.

[154] For details, see: http://www.mke.go.kr/language/eng/policy/Ipolicies.jsp.

East Asia's national energy strategies and systems are still generally very dependent on securing fossil fuel resources. This is indicative of the broader problem of how many of East Asia's core industrial structures, policies and practices remain high-carbon development geared. In Indonesia, for instance, the continued heavy subsidisation of fossil fuel prices make it difficult for the nation's geothermal sector to seriously compete in the electricity generation market. Furthermore, many countries persist with old industrial policies involving mass-scale ecological damage (e.g. Indonesia and Malaysia's palm oil industries) whilst rolling out low carbon development strategies (Gunningham 2011). Despite the March/April 2011 Fukushima disaster, many East Asian countries may still maintain their ambitious plans on nuclear power sector development, and moreover include nuclear under the policy rubric of green or clean energy. In sum, East Asia's 'modernist' industrialised development is likely to remain entrenched for some considerable time given the enormity of the structural change required to establish the broad foundations of low carbon development.

East Asia's developing countries especially face difficult challenges moving toward low carbon development given both the high proportionate GDP costs of replacing old or rudimentary equipment and infrastructure, and their relative lack of indigenous technological and innovatory capacity. Furthermore, there are domestic political pressures to meet more immediate socio-economic needs (e.g. poverty alleviation, provision of basic welfare and utility services) than to prioritise environmental-related goals. This being said, socio-economic and environmental problems are increasingly conflating in developing country regions. Many of East Asia's major cities (most notably China's) are subject to acute levels of pollution, causing chronic health problems on a mass scale and palpably deteriorating societal welfare. Even in relatively poor and authoritarian states there has been mounting civil unrest regarding pollution and a growing acknowledgement of problems caused by pursuing economic growth whatever the costs. For example, in August 2011 a protest from 12,000 inhabitants of Dalian compelled local authorities to relocate of a petrochemical plant, a reported 30,000 people in Haimen near Hong Kong marched in December 2011 against plans to construct a coal-fired power station, and thousands also took to the streets of Shifang in July 2012 to stop a new smelter plant being built.[155]

[155] *New York Times*, 5th July 2012.

Conclusion

By most calculations, East Asia is having more of an impact on global energy security and climate change than any other region, and the magnitude of this impact is widely expected to grow in years and decades to come. The expansion of renewable energy practice in East Asia is vital to addressing energy security and climate change challenges faced by the region as well as at the global level. This paper has examined the development of renewable energy sectors and policy in East Asian countries, and how this has become the basis of a formative 'new developmentalism', where the transformative objectives of low carbon development have been central to revitalised forms of state capacity. As we have seen, renewable energy and green energy more generally have become the focus of an emerging 'clean industry policy' paradigm that is embedded in recent strategic macro-development plans, such as Japan's New Growth Strategy, South Korea's Green Growth Strategy and the Strategic Emerging Industries component of China's 12th FYP.

However, it has been noted that low carbon development is not the only main objective of East Asia's new developmentalism. Its associated macro-development plans have too included the simultaneous promotion of high-carbon industrial activities, including conventional fossil fuel based sectors. Such apparent contradictions are likely to persist for some time given the structural dependencies of East Asian economies on high-carbon intensive industries for delivering material growth and prosperity, and that, like most others around the world, East Asian states have only just embarked on the very long transition to meaningful low carbon development, a process that most expect to take many decades if indeed it is to be achieved. In addition, East Asia's new developmentalism has tended generally to afford primacy to economistic objectives rather than environmentalism, to economic growth rather than sustainable development *per se*. This begs the question of just how far East Asia's new developmentalism marks a departure from the economic modernisation oriented policies of the past. It is perhaps better considered an intermediate concept through which *relatively lower* carbon development can be achieved over the medium-term. Moreover, achieving low carbon development requires deep social transformations, especially changing people's mindsets towards creating a green society that cannot be realised solely through a top-down policy process. The Fukushima incident of March/April 2011 may help provide greater societal and government support for renewables in East Asia in the shorter-term but this cannot be assured over the long-term. Nevertheless, by maintaining and improving their various forms of state capacity over time, East Asian states will be well positioned to sustain the significant recent advance of their renewable energy sectors and thereby further strengthen the low carbon development orientation of their new industrial policies, macro-development plans and strategic economic thinking. The deeper diffusion of renewable energy technologies in every element of society is also a critical socio-technical process to secure lower carbon outcomes, in East Asia and elsewhere.

Bibliography

Agency for Natural Resources and Energy (ANRE). 2010. *The Strategic Energy Plan of Japan: 2010 Revision*. ARNE, Tokyo, Japan.

Amsden, A. 1989 *Asia's Next Giant: South Korea and Late Industrialisation*. Oxford University Press, Oxford, UK.

China Council for International Co-operation on Environment and Development (CCICED). 2010. China's Pathway Towards a Low Carbon Economy. CCICED, Burnaby, Canada.

Energy Information Administration. 2010. *International Energy Outlook 2010*. EIA, Washington, DC, USA.

European Photovoltaic Industry Association (EPIA) 2012. *Global Market Outlook for Photovoltaics until 2016*. EPIA, Brussels, Belgium.

Gills, B. (ed.). 2010. *Globalisation in Crisis*. Routledge, London, UK.

Gunningham, N. 2011. 'Energy Governance in Asia: Beyond the Market', *East Asia Forum Quarterly*, 3(1): 29-30.

Hayashi, S. 2010. 'The Developmental State in the Era of Globalisation: Beyond the Northeast Asian Model of Political Economy', *The Pacific Review*, 23(1): 45-69.

International Energy Agency (IEA). 2012. *Global Renewable Energy: Policy and Measures Database*, IEA, Paris, France. Available at: http://www.iea.org/textbase/pm/?mode=re.

Johnson, C. 1982. *MITI and the Japanese Miracle*. Stanford University Press, Chicago, IL, USA.

Jung, T.Y. & Ahn, J.E. 2010. 'Sowing the Seeds for Green Growth in Korea', *East Asia Forum*, 13 December 2010.

Karagiannis, N. & Madjd-Sadjadi, Z. 2007. *Modern State Intervention in the Era of Globalisation*. Edward Elgar, Cheltenham, UK.

Korea Environmental Industry and Technology Institute (KEITI). 2009. *Korea's Green Growth Vision and Eco-Innovation Policies*. KEITI, Seoul, South Korea.

Kwon, H.S. 2010. Research on the Way to Foster Leading New and Renewable Energy Industry on Greater Sphere Economic Area. *KEEI Research Paper, 31 December*. Korea Energy Economic Institute, Seoul, South Korea.

Lee. B., Iliev, I. & Preston, F. 2009. *Who Owns Our Low Carbon Future?: Intellectual Property and Energy Technologies*. Chatham House, London, UK.

Lim, H. 2010. The Transformation of the Developmental State and Economic Reform in Korea, *Journal of Contemporary Asia*, 40(2): 188-210.

Martinot, E. & Li, J.F. 2010. Renewable Energy Policy Update for China. *Renewable Energy World*, 21 July 2010. Available at: http://www.renewableenergyworld.com/rea/news/article/2010/07/renewable-energy-policy-update-for-china.

Ministry of Economy, Trade and Industry (METI). 2010a. *The History of Japan's Industrial Policies*. METI, Tokyo, Japan.

Ministry of Economy, Trade and Industry (METI). 2010b. *The New Growth Strategy: Blueprint for Revitalising Japan*. METI, Tokyo, Japan.

Ministry of Economy, Trade and Industry (METI). 2010c. *The Industrial Structure Vision 2010 (Outline)*. METI, Tokyo, Japan.

Moon, T.H. 2010. Green Growth Policy in the Republic of Korea: Its Promises and Pitfalls. *Korea Observer*, 41(3): 379-414.

National Development and Reform Commission (NDRC). 2007a *Medium and Long-Term Development Plan for Renewable Energy in China*. NDRC, Beijing, China.

National Development and Reform Commission (NDRC). 2007b. *China's National Climate Change Programme*. NDRC, Beijing, China.

National Development and Reform Commission (NDRC). 2011. *China's 12th Five-Year Plan for Economic and Social Development*. NDRC, Beijing, China.

Noland, M. 2007. From Industrial Policy to Innovation Policy: Japan's Pursuit of Competitive Advantage. *Asian Economic Policy Review* 2(2): 251-268.

Olz, S. & Beerepoot, M. 2010. *Deploying Renewables in Southeast Asia*. IEA, Paris:, France.

Pirie, I. 2008. *The Korean Developmental State: From Dirigisme To Neo-Liberalism*. Routledge, London, UK.

Radice, H. 2008. The Developmental State under Global Neoliberalism. *Third World Quarterly*, 29(6): 1153-1174.

REN21. 2006. *Renewables 2006 Global Status Report*. REN21 Secretariat, Paris, France.

REN21. 2013. *Renewables 2013 Global Status Report*. REN21 Secretariat, Paris, France.

Renewable Energy and Energy Efficiency Partnership (REEEP). 2012. *Policy and Regulation Review Database*. REEEP, Vienna, Austria. Available at: http://www.reeep.org/9353/policy-database.htm.

Rodan, G. 1989. *The Political Economy of Singapore's Industrialisation: National State and International Capital*. MacMillan, Basingstoke, UK.

Stubbs, R. 2009. What ever Happened to the East Asian Developmental State? The Unfolding Debate. *The Pacific Review* 22(1): 1-22.

Suliman, O. (ed.). 1998. *China's Transition to a Socialist Market Economy*. Praeger, London, UK.

Wang, T. & Watson, J. 2009. *China's Energy Transition: Pathways for Low Carbon Development*. Sussex Energy Group and Tyndall Centre for Climate Change Research, Brighton, UK.

Weiss, L. 1998. *The Myth of the Powerless State: Governing the Economy in a Global Era*. Polity, London, UK.

Weiss, L. 2010. The State in the Economy: Neoliberal or Neoactivist? In Campbell, J., Crouch, C., Hull Kristensen, P., O. K. Pedersen, and R. Whitley (eds.). *Oxford Handbook of Comparative Institutional Analysis*. Oxford University Press, Oxford, UK.

Woo-Cumings, M. (ed.). 1999. *The Developmental State*. Cornell University Press, Ithaca, NY, USA.

Contributors

Camilla Brautaset: Associate Professor in modern economic history at the Department of Archaeology, History, Cultural Studies and Religion, University of Bergen, Norway. Most recent publication: Fortellinger og forklaringer i økonomisk historie (with L.F. Øksendal). In: Heiret, J., Ryymin, T. & Skålevåg, A. (eds.). *Fortalt fortid. Norsk historieskriving etter 1970*. Pax Forlag, Oslo, Norway: 292-319, 2013.
E-mail: camilla.brautaset@ahkr.uib.no

Scott Bremer: Researcher at the Centre for the Study of the Sciences and the Humanities, University of Bergen, Norway. Most recent publication: Exploring the science-policy interface for integrated coastal management in New Zealand. *Ocean and Coastal Management*, 84: 107-118, 2013.
E-mail: scott.bremer@svt.uib.no

Peter Buckley: Professor of International Business, Centre for International Business, Leeds University Business School, UK. Most recent publication: Adam Smith's theory of knowledge and international business theory and practice. *Journal of International Business Studies*, 2013. Available at: http://dx.doi.org/10.1057/jibs.2013.44
E-mail: p.j.buckley@lubs.leeds.ac.uk

Christopher M. Dent: Professor of East Asia's International Political Economy, University of Leeds, UK. Most recent book publication: *The Asia-Pacific, regionalism and the global system*. Edward Elgar, Cheltenham, UK, 2012.
E-mail: c.m.dent@leeds.ac.uk

Peter Enderwick: Head of the Centre for International Business, Auckland University of Technology, New Zealand. Most recent publication: Country linkages and firm internationalisation: Indian MNEs within economic-political alliances of nations. In: Cook, G. & Johns, J. (eds.). *The changing geography of international business: strategy, performance and institutional change*. Palgrave MacMillan, Basingstoke, UK, 2013.
E-mail: peter.enderwick@aut.ac.nz

Nicolas Forsans: Associate Professor in International Strategic Management, Centre for International Business, University of Leeds, UK. Most recent publication: Country linkages and firm internationalisation: Indian MNEs within economic-political alliances of nations. In: Cook, G. & Johns, J. (eds.). *The changing geography of international business: strategy, performance and institutional change*. Palgrave MacMillan, Basingstoke, UK, 2013.
E-mail: n.forsans@lubs.leeds.ac.uk

Michael Hertzberg: PhD Candidate at the Chr. Michelsen Institute, Bergen, Norway & the Department of Archaeology, History, Cultural Studies and Religion, University of Bergen, Norway. Most recent publication: Med slagord som våpen: Politiske munker og borgerkrigen på Sri Lanka. *DIN: Religionsvitenskapelig tidsskrift,* 3-4: 48-59, 2011.
E-mail: michael.hertzberg@cmi.no

Lewis Husain: Post-doctoral Research Fellow, Institute of East Asian Studies, University of Duisburg-Essen, Germany. Most recent publication: *Implementing and innovating: local governments in the development of China's New Cooperative Medical Scheme.* Unpublished PhD thesis, University of Leeds, UK, 2013.
E-mail: l.husain@wreac.rog

Janne Cecilie Johansen: PhD candidate at the Centre for the Study of the Sciences and the Humanities, University of Bergen, Norway. Most recent publication: Grenseoverskridende sosiologi: et intervju med vinneren av Holbergprisen 2012, Manuel Castell. *Sosiologisk Tidsskrift*, 20(4): 361-371, 2012.
E-mail: janne.c.johansen@svt.uib.no

Matthias Kaiser: Professor in Philosophy of Science at the Centre for the Study of the Sciences and the Humanities, University of Bergen, Norway. Most recent publication: Precaution or prudent vigilance as guiding the path to global food security? In: Röcklinsberg, H. & Sandin, P. (eds.). *The ethics of consumption - The citizen, the market and the law*. Wageningen Academic Publishers, Wageningen, the Netherlands, 2013.
E-mail: matthias.kaiser@svt.uib.no

Ole Jacob Madsen: Associate Professor at the Department of Psychology, University of Oslo, Norway. Most recent publication: Me at my best: therapeutic ideals in Norwegian women's magazines. *Communication, Culture & Critique*, 5: 20-37, 2012.
E-mail: o.j.madsen@psykologi.uio.no

Olga Medvedeva: PhD candidate at the Department of Archaeology, History, Cultural Studies and Religion, University of Bergen, Norway. Most recent publication: Lu Xun in the rhetoric of the Sino-Soviet split: a view from contemporary Russia. *Frontiers of Literary Studies in China*, 7(3): 483-493, 2013.
E-mail: olga.medvedeva@ahkr.uib.no

Surender Munjal: Lecturer in International Business & Director Research, James E. Lynch India and South Asia Business Centre, Centre for International Business, University of Leeds, UK. Most recent publication: Country linkages and firm internationalisation: Indian MNEs within economic-political alliances of nations. In: Cook, G. & Johns, J. (eds.). *The changing geography of international business: strategy, performance and institutional change.* Palgrave MacMillan, Basingstoke, UK, 2013.
E-mail: s.munjal@lubs.leeds.ac.uk

Simen A. Øyen: PhD Candidate at the Centre for the Study of the Sciences and the Humanities, University of Bergen, Norway. Most recent publication: *Humanioras fremtid.* Cappelen Akademiske Forlag, Oslo, Norway, 2011.
E-mail: simen.oyen@svt.uib.no

Michael Parnwell: Distinguised Guest Professor at the Faculty of Hospitality and Tourism Management, Macau University of Science and Technology, China & Emeritus Professor of South East Asian Development at the University of Leeds, UK. Most recent publication: Whale worship and tourism development in the Hoi An-Da Nang corridor, Viet Nam. *South East Asia Research,* 21(3): 475-496, 2013.
E-mail: m.j.g.parnwell@leeds.ac.uk

Karina Hestad Skeie: Associate Professor in Intercultural Studies at NLA University College, Bergen, Norway. Most recent publication: Building God's Kingdom. Norwegian missionaries in highland Madagascar 1866-1903. *Studies in Christian Mission* 42. Brill, Leiden, The Netherlands, 2012.
E-mail: karinahestad.skeie@nla.no

Arne Sveinson Haugen: Researcher at the Centre for the Study of the Sciences and the Humanities, University of Bergen, Norway. Most recent publication: Mapping core values and ethical principles for livelihoods in Asia. In: Potthast, T. & Meisch, S. (eds.). *Climate change and sustainable development. Ethical perspectives on land use and food production.* Wageningen Academic Publishers, Wageningen, The Netherlands: 419-424, 2011.
E-mail: arne.sveinson@svt.uib.no

Adam Tyson: Lecturer in Southeast Asian Politics at the School of Politics and International Studies, University of Leeds, UK. Most recent publication: Vigilantism and violence in decentralized Indonesia: the case of Lombok. *Critical Asian Studies,* 45(2): 201-230, 2013.
E-mail: a.d.tyson@leeds.ac.uk

Index

P

Q

R

S

Y

Z

www.ingramcontent.com/pod-product-compliance
Lightning Source LLC
Chambersburg PA
CBHW061207240926
51846CB00006B/1143